COOL SHOP

酷店

李亮亮 胡嘉思 编

辽宁科学技术出版社

目录
Contents

Ariel Rojo

The destiny of establishments is uncertain. The shopping habits find each day a new paradigm that become quickly in new custom; Internet, Telemarketing or even the postal service are all very profitable alternatives for not having to invert in physical spaces, however, why do we still need shops?

At last, none of the mentioned alternatives substitute the complete experience of going to a store. Observing the details of an item, watching it from all angles, feeling the softness of fabric, being able to try on the clothes, smelling the bread coming out from the oven, falling in love with someone you saw in the music shop; these are some of the experiences that these spaces offer us beyond the practice of just buying the product. Being inside, going outside, up and down, walking around, finding, and living in a space that send us to another world; the world inside the mind of a designer, or at least to another place in this world.

Our society needs objects, architecture and urbanism with more spiritual content, searching for the essence with new experiences. Today a designed shop can certainly become a piece of contemporary art.

The exterior design of a shop concerning its urban surroundings is going to be a theme of great interest in the upcoming years. Besides the shop display windows and the signs that are now still used in traditional ways; new resources are going to be experimented to call the attention of clients. They will offer different situations and new norms for urban spaces to let façade designs more audacious. Nevertheless we will always have to be capable of compensating its presence with the context.

Mexico City with all its "architectural liberties" is undoubtedly a show of plenty different styles. It is always difficult to conciliate so many architectural proposals. Commercially speaking one can appreciate an architecture and urban phenomenon with the

presence of so many brands. Some of them are placed in a harmonious way and many others in an invasive way.

It is possible to design an advertisement that appears in a subtle way, just insinuating or suggesting the related desires to the product or service. It is going to be the future designers´ job to achieve the correct balance between information, publicity and life in the metropolis.

1. How can you deal with the clients or their different opinions between your design and client's demand?

As part of the ADN of a project, the necessities of the client have to be included to achieve the project in the most harmonious way. The best client is the one that is open to new ideas and that is looking for the same things as you in design terms.

2.Your ultimate shop project...

Club Med for Mexico City, the corporative offices and the boutique.

3. You wish you'd designed…

A very sexy motel

4. Your favourite shops are…

Liberty (London)

建筑的命运毫无确定性。人们每天购物的习惯都不相同，这已经迅速成为新的风尚。网上购物、电视购物和邮购，这些购物方式都十分便利，人们无需再去逛实体商店。但是，我们为什么仍然需要商店?

因为在上述购物方式中，没有一个会让人有逛商店的感觉。观察商品的全部细节，从各个角度欣赏它，感受布料柔软的触感，试穿那些衣服，闻到烤箱里传出来的面包香味，在音像店邂逅某人并坠入爱河。商店带给我们的体验不止是买东西。进店再出来，上楼下楼，随意逛逛，寻找喜爱的商品，融入店内的另一个世界（设计师脑海之中的世界），或至少是这个世界上的另一个地方。

我们的社会需要赋予物体、建筑和城市规划更多的精神内容，寻求新体验的内涵。如今，精心设计的店面必定会成为当代艺术的一部分。

一家商店的外观设计关系到其所在地区的周边环境，也将在未来几年中，成为极具吸引力的主题。除了橱窗和现在仍然用于传统设计中的标志以外，还会加入吸引顾客目光的新元素。如在市区中以不同的定位和新的标准进行选址，把外观设计得更加大胆。不过，我们也会使它与周围环境保持和谐。

“建筑自由化”的墨西哥城内无疑展现了大量不同的建筑风格。但太多的建筑风格始终无法调和。从商业角度来说，你能欣赏这里的建筑和城中大量涌现的品牌。其中有的品牌选址相对和谐，而有的则带有侵略性。

在广告设计中，会以一种微妙的方式来对某种产品或服务进行暗示或启发人们相关的欲望。这也将成为未来设计师们的工作，以达到大都市中信息、宣传和生活之间的正确平衡。

1.你会如何处理与客户之间的关系，以及他们对于你的设计要求的不同意见?
作为ADN的一部分，客户必须与该项目达成最和谐的共识。最佳客户要有开阔新颖的思路，他所期待的东西恰好与你的设计理念一致。

2.你最满意的店面设计作品是……
墨西哥城的地中海俱乐部，是我们的合作单位和精品店。

3.你最想设计的是……
一幢非常性感的汽车旅馆。

4.你最喜爱的店面是……
伦敦的“自由”。

BART LENS

In recent years there has developed a trend of an increased specificity of concept for each store – Nike has its specific concept, coffee stores have their own clearly defined and differing concepts. As a reflection of this, the features of these shops form part of the concept in various, complimentary ways, and support a sort of "lifestyle" – and increasingly this typology is replacing the typical concept of just a few minutes of looking and buying, with an experience of browsing, reading, sitting, eating; in fact something like another way of life, one particular to this store.

For what lies in the future, I am sure that shop designers will be concerned increasingly with interaction; you might be walking past a clothes store, and a little chip in your Levi jeans allows the shop you are passing to "know" this, and in some manner proposes you a similar or related product or provides you with some information on a future product.

In the same line of thought, it is the future clients with which shops must be able to communicate and interact; it will no longer simply be a case of product first, then communication, rather the other way around. As an abstract example, water does not need to be re-invented, rather the thing which makes the critical difference; it is the way by which it is served.

For example, a pharmacy theoretically does not have to do this, as people have a "need" of what it provides, yet what effects the success of this particular pharmacy is the environment in which it provides this "need" to its clients. As noted previously, people increasingly desire lifestyle choices, so much so that this desire is a kind of "need" – a sort of peg to hang yourself on, for whatever reason, emotion, or need. Yet the very best shops and stores don't even need to search for this – it is already part of what they do and what they are.

As a sort of conclusion, the communication between client and designer is relevant in this discussion – indeed it must be exactly that – a two-sided conversation which allows wish to be translated into form and function, and vice versa; it is most often through this creative process from which the best designs emerge, those that provide this "total experience".

1.How do you deal with the clients or their different opinions between your design and client's demand?

Our clients come to us because they are already familiar with our previous work, and therefore convincing them of "what they want" is not so necessary, as they already know what we provide. The principle discussion therefore, and as is often the case in architecture and design, is cost. It is never the case that the client controls the design process; they have their demands and wishes in terms of functionality and use etc, and we take these and shape them into concepts, giving them form, producing an aesthetic vision particular and unique to the project in question.

2.Your ultimate shop project...

In Hasselt we had the opportunity to create a design shop from the old post office of the town. Not only was this a special project in itself, but the project encompassed all scales of design; we produced the shop concept and interior, and we also restored the building and reconfigured the street itself, allowing us to work in a holistic manner, creating a "city experience". It is this sort of "total concept" which is so interesting to produce, working in-between the smallest detail and the urban whole.

3.You wish you'd designed...

The Hempel Hotel by Anushka Hempel was a design I experienced some ten years ago and it was a moment where I thought to myself "Imagine I could make this" – in the sense that it was so well calibrated; from the small to the large scale, again a holistic approach to design, one with a total experience which had been articulated at every level, and, what a level!

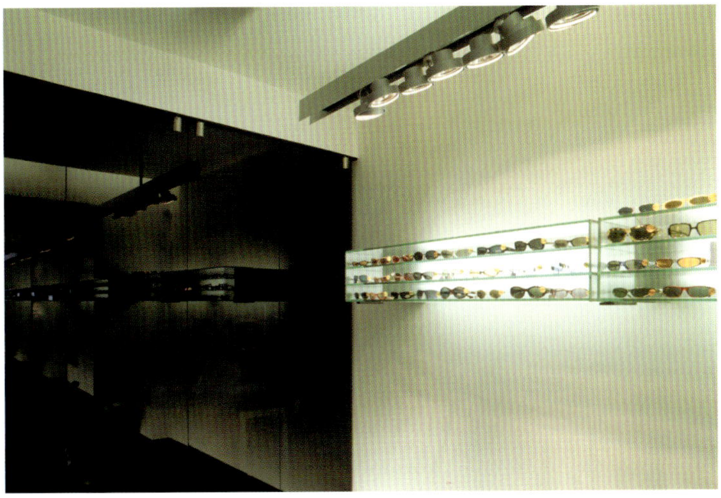

4.Your favourite three shops are…

In Milan, "Corso Como 10" is a place where one can stay for hours; it is a shop, a library and there are clothes, design products. In fact it seems so disorganised, almost messy, yet there is clearly an order to the apparent chaos; an underlying organisation and structure. Another favourite is a bread/coffee shop nearby to Corso Como 10, designed by Claudio Sylvestrin. Finally, in New York the department store "Nike Town" is memorable for the different levels of products, but not only that, but different experiences; in each area you can smell and hear things related to the products themselves; in the swimming accessory level you can hear water, smell water. It is the sort of store concept which is almost unimaginable here in Belgium.

近年来，店铺理念越来越个性化——耐克有其特别的理念，咖啡店也有界定清晰、各不相同的理念。于是，商店的理念五花八门，互为补充，支持着一种“生活方式”。个性化越来越多地取代了有代表性的商店理念，你只需几分钟就可以找到需要买的商品，体验浏览、阅读、闲坐、饮食的乐趣。这实际上是另一种生活方式，是这家店的特殊之处。

我认为，未来的发展趋势是商店设计者会越来越注重互动。你从一家服装店门前走过，你牛仔裤上的一块小芯片让这家店“感知”到你就在附近，它会以某种方式向你推荐类似或相关的商品，还能为你提供即将发布的商品信息。

同样，商店必须与“未来”的顾客沟通互动，商店不再只是简单地先展示商品，再跟顾客沟通，而是需要有别的方式。举个例子，水是不需要被重新发明的，但它在不同状态下能起到完全不同的作用。

比方说，药房就不需要做这些事，人们有“需求”时才会去药房。于是影响药房的成功因素在于它是在怎样的环境中为顾客提供“需求”。如前所述，人们越来越渴望选择生活方式，渴望就是一种“需求”，它把你和理由、情绪或需要联系在一起。然而，最好的商店甚至不用去寻找顾客的需求，它已经存在于商店之中了。

结论：客户和设计师之间的沟通十分重要，双边对话有利于把愿望转化为形式与功能，反之亦然。这种沟通最常出现在最佳设计的创作过程中，正是这些过程构成了这次设计的“全部体验”。

1.你如何处理客户及他们对你的设计要求提出的不同意见?

来找我们的客户一般都对我们之前的作品有所了解，他们已经知道我们会做什么，因此无需再确认他们想要的东西。建筑设计之中，我们经常会探讨费用问题。客户从不干预设计过程，我们把他们在项目功能、用途等方面的愿望要求融入到设计理念之中，化为有形的东西，做得既符合美学视角又独一无二。

2.你的最佳商店设计作品是……

在哈塞特，我们设计了一个商店，它的旧址是邮局。这个项目本身就很特别，而且它还涵盖了设计工作的所有方面。我们为它做了理念和内部装修，还修复了整个楼体，并对它所在的街道也做了修整。我们用整体的设计方式，塑造了一个“城市体验”。创造一个“整体理念”很有趣，在小细节和大都市之间做文章。

3.你最想设计的是……

十多年前我为阿努什卡·亨普尔设计了亨普尔酒店，当时，我在心里对自己说，“想象你能做好”，从这点看来，作品完成得相当不错。从小处到大处，又是一次整体的设计，每一个层面都是我的全部经验的结晶。

4.你最喜欢的三家商店是……

在米兰，“科索科摩10”会让你流连忘返。它是商店和图书馆，有服装和各种设计品。那其实有点凌乱，却乱中有序，组织秩序潜藏其中。我喜欢的另一家店是“科索科摩10”旁边的面包咖啡店，由克劳迪奥·希尔维斯金设计。还有纽约的“九城”百货公司，让人印象深刻的是不同档次的商品，不只这些，还有不一样的购物体验。在每一区你都能闻到、听到产品本身发出的信息，在游泳产品区你就能听到水的声音，闻到水的气息。那里的店铺理念在比利时根本无法想象。

Emmanuelle Moureaux

Internet shopping, TV shopping, mobile phone shopping, the way of buying has been revolutionised these last years by the development of the "Home shopping". You don't need more to go to shops; shops from all over the world come to you. Although this new shopping style becomes more and more convenient, cheaper and cheaper, why are we still enjoying the "shop shopping"?

In the past, shops were not only a place for buying, but a really place for communication. There was a really conversation between the customer and the seller, or between customers too. We didn' t only buy something; we enjoyed these human relations.

Then, shops, located in shopping streets, moved to shopping centres, or multi-purposes buildings. The proximity shopping tends to disappear for a targeted shopping. We don't more choose shops in function of their location or the persons who work in, but we select them by their image. The design became an unavoidable factor, as important as the products.

In Tokyo, new shops open all the time; only few years old shops are renewed in an incredible rhythm. These new shops are not only a place for buying, but they recreate a particular universe. Thus, going shopping is like making a trip, a dream which project you in a no-ordinary world. Shopping became leisure. As shops are designed more as a show-room than a meeting place, shopping remains an individual trip.

A shop must not be only a business place but needs to supply something more. It is very important for me the space design gives emotions to people, and spurs naturally to communicate. Create a shop space like that implies to find a good balance between all its elements: the space and its design, the goods and their display, the signs, and people. A lot of new shops seem forget people.

Recent shops have identical features. They recreate a fantastic, a futuristic, or a luxurious universe, in any case a no-ordinary one. Because people realise now the importance of the space design, its budget becomes higher and higher and luxurious shops are no rare anymore. These new shops, very opened to the exterior, are characterised by fluid lines, brilliant and smooth textures, and monotone colours. Goods are displayed like in a museum or a show-room. Because the general design image remains quite neutral, all these shops look identical at first glance.

But "cool" doesn't only mean cold or fashionable. A cool shop is a space whose design you ever seen before, which gives special feelings to its visitors or surprise them.

Although a prodigious development of new technologies, there are very few changes in the fields of architecture and interior design, in comparison to cars, audio-visual or computer industry. Living, dining, kitchen, bedrooms, the typical plan of a house has not been changed for several decades and the evolution of the architectural space, interior space and furniture remains very slow. Appearances change, but concepts or fundaments not. Of course, it is the same situation for shops.

The trend of the future shops may be a return to an environment nearer to people, a change from no-ordinary to ordinary spaces. Shops may be less luxurious, less short-lived. While people remain passive in present shops, the future shops may create a new kind of communication, by interactivity for example and may become a new place of meeting.

I am designing a lot of shops in Japan, like cooking schools (ABC Cooking Studio), lunch boxes shop (Lunch Vacation), hair and beauty salon (arp hills), show-room (CS Design Center) and others.

The most important thing for me is to design new concepts, new atmospheres which will give emotions to people who visit or use the space. I create spaces using colours, or it is better to say the colours I use create the space. I use colours as three-dimensional items, like layers, in order to create spaces, not as an element applied

after in order to complete or decorate the space. I am living and working in Tokyo for 12 years, a city which inspires me a lot. Tokyo is very colorful, contrary to the grey monotonous European cities, built in stones. The buildings themselves are not coloured, but a lot of non- architecture elements like vending machines or signboards give colours to Japanese cities. In the past, there were a lot of colours in the Japanese daily life. People wear colourful kimonos and vivid red colour was often used for the construction of temples. The recent buildings or interiors are monotonous, and white, black or grey are very fashionable colours in Japan. I try to express in my projects some Japanese essence, like colours, which has disappeared or has tendency to disappear in the contemporary designs. I try to reintroduce colours in a new way in the daily life.

Transparency is another key-point in all my projects. I don't use walls to divide the space but create flexible and fluid spaces, which communicate and breath, with ambiguous limits.
"Colours" and "transparency" are expressed in a concept I develop in all my projects, the concept of "Shikiri", a made-up word which literally means "dividing space with colours" in English. "Shikiri" is a colourful partition series, inspired by the Japanese traditional sliding paper screens or wood partitions, which have almost disappeared now.

By creating and designing a shop space with colours, I try to create a strong visual image by the interior design. The shop spaces I design catch eyes and remain in memory. Apart the visual appearance, people are very important in my projects. I not only try to give emotions to users, but try to create communication by rethinking all the uses in the shop. It is a parallel work, on invisible parts like use and users, and visible parts like the interior visual design. "Cool shops" mean "Colourful Emotional Shops" for me.

网上购物、电视购物、手机购物……近年来，随着“家庭购物”的发展，购物方式发生了重大的变革。你无需走入商店，全世界的商店就在你的面前。既然这种新的购物方式越来越便捷，价格也越来越便宜，我们为什么还喜欢到商店里买东西呢?

过去，商店不仅是买东西的地方，更是个交流的地方。顾客和卖家之间或顾客之间都存在着真正的对话。我们不光买了东西，还享受着人与人之间的关系。

后来，商业街旁的商店都搬进了购物中心，或多功能大楼。邻近的购物区消失了，我们逛街也变得有针对性。我们不会根据地点和工作人员来选择商店，而是选择样子漂亮的店面。设计变成了不可缺少的因素，有着和商品同等重要的地位。

在东京，时时都有新店开业，没开几年的旧店也以惊人的速度更新换代。这些新店铺并不只是买东西的地方，它们构成了一个特殊的世界。逛街就像旅游，像一个梦想，把你带入这个不平凡的世界。购物成为一种休闲。当看到眼前的商店装饰得像展示厅般琳琅满目，而不是简单的聚会场所，购物依旧是一次旅行。

商店不光是商业场所，还要有更多的东西。空间设计给人们的感觉对我十分重要，人们交流的欲望会自然而然的增加。打造一个商店意味着在商品的所有元素之间获得平衡：空间与设计、商品与展示品、标志、面向人群。很多新商店好像忘记了人这一元素。

相邻的店铺都有着类似的特点。他们建立了一个充满未来幻想的豪华世界，拒绝平庸。人们意识到了空间设计的重要性，这方面的预算也越来越高，豪华的商店不再是凤毛麟角。这些新商店外部都装潢

1)How can you deal with the clients or their different opinions between your design and your client's demand?

In general my clients give me carte blanche for the design. I can imagine freely how to create a unique space which will give emotions and special feelings to its future users. But it doesn't mean that I can decide everything. Especially for the projects I design, like cooking studios, cafes, beauty salons and others, there are always restrictions -sizes, material, cost, schedule- and a lot of practical requests I have to respond to. When a project starts, I discuss a lot with my client in order to understand what is important for him about practical use, space operation etc..., what is absolutely necessary and what is not. I give priorities to these different elements and choose to integrate them as much as possible in the project if I think they are necessary, or decide not if I think they will weaken the general concept. By discussing a lot with my clients, I can understand their dreams, I can explain them the design concept, so they can understand my vision of the project. I think the most important thing is to communicate as much as possible.

2)Your ultimate shop project...

I would like to design totally a shop, from the brand concept, packaging, graphics to the interior design. In general when you design a shop, the brand concept is already decided, the brand logo already designed, the packaging too. Thus the interior designer must create an atmosphere which fits perfectly to the brand concept. That was the case when I designed Lunch Vacation: Although Lunch Vacation was a new brand, the logo and its colours was already decided when I start the project because the branding image, graphic design, interior design, all these works were very split and entrusted to different persons. But I think interior design, graphics, packaging design, and the brand concept must be in total coherence so I dream to design totally a shop from the brand concept. I love stationery, I love chocolate, and I am a French architect living in Tokyo so it would be fantastic if I had the opportunity to design a chocolate shop or a stationery shop in Europe, from the brand concept, packaging, graphics to the interior design. Of course I
love colours, so I would design a very colourful shop, in which the products and colours match perfectly together.

3)You wish you'd designed...

I discovered by chance a small chocolate shop, called XOCOA in the old Barcelona. I think this brand is famous in Spain but I didn't know it before. The brand concept is "the new experience of chocolate", so this not common shop tries to give a new vision of the chocolate. The interior design is very simple. The walls and the ceiling are painted in dark pink, and a lot of bars of chocolate are displayed on a white wall. They have very beautiful and colourful packages which highlight them and give a lot of colours to the shop. Each package is fantastic and visiting the shop is an amazing visual experience. I really love it. The century-old family enterprise decided to modernise and give a new image to their brand so they asked graphic designers to redesign their packaging and create new products. It is a very interesting project and I wish I had designed a project like this because it is a real work on the total brand image. It is a very small shop but it is a real pleasure to visit it.

得十分张扬，饰有流线和光影流动的织物，使用统一色系。它们像博物馆和展览厅里那样展示商品。一般的店面形象都没什么特别之处，因此这些店铺乍看上去都差不多。

但“酷”不仅仅意味着冷漠和时髦。酷店的设计是你之前从未见过的，给来访者特别的感受或惊喜。

尽管新技术的发展十分惊人，但是和汽车、视听产品或计算机业相比，建筑和室内设计方面的变化却少得可怜。起居室、餐厅、厨房、卧室，房屋的典型格局几十年来一成不变，建筑空间、内部空间和家具的变革也十分缓慢。

外形虽然变了，但概念或本质并没有变。当然，商店也是同样的情况。

未来商店的趋势是重返人们的身边，并从不平凡走向平凡。商店可能不再那么豪华，寿命也长些。现在，人们在店内还是被动、顺从的角色，未来会出现买卖互动等新的交流方式，商店可能会变成新的会面场所。

我在日本设计了很多店铺，像烹饪学校（ABC烹调工作室）、便当店（假日午餐）、美容美发沙龙（阿普颠峰）、展示厅（CS设计中心）等等。

对我来说，最重要的是设计出全新的概念和氛围，把来访者和店内工作人员的情绪都带动起来。我用色彩塑造空间，或是说我运用的色彩塑造了空间。我对颜色做分层处理，做成三维的效果来创造空间，而不是把它当成完成装饰的一个元素。我在东京生活、工作了12年，这个城市给了我很多灵感。东京非常丰富多彩，和用单调的灰色石头堆砌出来的欧洲城市截然不同。东京的建筑本身是没有颜色的，但自动售货机和招牌等非建筑元素把这座城市装点得多姿多彩。过去，日本人的日常生活中也充满了色彩。人们都穿着彩色的和服，庙宇上也有生动的红色颜料。近年来，建筑内外都开始流行起白、黑、灰的色调。我试着从我的作品中表达日本的本质，使用当代设计中已经消失或即将消失的颜色。我要以全新的方式把颜色重新引入日常生活。

通透感是我作品的另一个要素。我不用墙来分割空间，我塑造的空间柔韧流动，能感受到在微妙局限中的交流和气息。

在我的作品中，颜色和通透感表达了“Shikiri”的概念，这是一个再造的词，意思是“以色彩分割空间”。“Shikiri”是一个色彩分割的系列，灵感来源于日本传统的滑动纸帘和木质屏风，这些东西在现在已经不多见了。

通过色彩来创造店铺空间，我试着在店内作出强烈的视觉效果。我设计的商店都十分吸引人的眼球，令人印象深刻。除了这些表面的东西，人在我的作品中也很重要。我不仅要带动商店使用者的情绪，还要设想到商店的全部用途，试着创造沟通。这不是一件容易事，要考虑到无形的用途和使用者，还要兼顾有形的店内设计。“酷店”对我来说就是“色彩丰富的感动之店”。

1.你如何处理与客户之间的关系，以及他们对你的设计要求提出的不同意见?
一般来说，我的客户都让我全权负责设计工作。我可以自由想象怎样塑造一个独特的空间，给它未来的使用者带来感动和特别的感觉。但这并不表示我可以决定一切。特别是我设计过的项目，烹调演播室、咖啡馆、美发沙龙等，在面积、材料、成本、日程安排上总有一些限制，还有很多实际要求我必须去迎合。项目开始时，我常常跟客户探讨问题，以了解他在实际操作和空间管理中更加注重哪些方面，哪些是十分重要的，哪些可以忽略。我优先考虑这些不同的元素，必要的或是我认为不会削弱大原则的部分会在项目中得到完善。通过和客户的探讨，我了解了他们的梦想，再向他们讲解设计理念，这样他们就能理解我的设计目标。我认为最重要的是尽可能多地沟通。

2.你最满意的商店设计作品……
我喜欢设计完整的商店，从品牌概念、包装、商标到内部装修。一般来说，设计一间店铺的时候，品牌概念已经有了，商标和包装也已经设计好了。这样室内设计师创造的店内氛围一定要与品牌概念相吻合。我设计“假日午餐”时就遇到过这种情况：“假日午餐”虽然是一个新品牌，但在我接手设计工作时，商标和色调都已敲定，因为设计商标、图案、店内造型的这些工作是由不同的人做的，所以风格不太和谐。我认为室内设计、商标和包装设计必须与品牌概念保持统一性，所以我最想从品牌概念着手，设计整个商店。我爱文具，我爱巧克力，我是一个在东京生活的法国建筑师，如果我有机会在欧洲设计一个巧克力商店或文具店，从品牌概念、包装、商标到室内装修全部包办，那就太完美了。

3.你最想设计的是……
一个偶然的机会，我在古老的巴塞罗那发现了一家小巧克力店，名叫XOCOA。这个牌子在西班牙好像很出名，但我以前却从没听说过。它的品牌概念是“巧克力的新体验”，这种把巧克力赋予新视角的店铺并不多见。店内设计十分简单。墙壁和天花板漆成了深粉色，在一面白色的墙上展示着很多巧克力棒。它们太漂亮了，在彩色的包装纸下特别醒目，也为小店增色不少。每个包装都如梦幻般美丽，逛这家商店就是一次奇妙的视觉体验。我真的太喜欢它了。这个有着百年历史的家族企业决定现代化，在他们的品牌中加入新的形象，于是他们让商标设计师重新设计了包装并推出了新产品。这是个非常有趣的项目，我希望自己能设计一个像这样的店，这是真正的设计完整的品牌形象。这家店很小，但它真的令人很愉快。

Greg Natale

Our approach to retail has always been the same as our approach to any form of design, to find exciting and original design solutions for our clients. This quality is particularly important when dealing with retail. Given that the lifespan of a retail space is constantly subject to change, being on the fore front of trends is of the utmost importance. Like all good design the aim is to create a useable space that can stretch the design limits and bring something new and exciting into the mix of your everyday space. We accomplish this by using strong colour palettes, bold graphics and a sense of playfulness juxtaposed with superb quality and a passion to do something that produces that "wow factor," leaving the client more than happy with the results.

In Retail as with all our design we are always searching for ways to produce that vibe that not only defines our client but has the user intrigued for more. With the way stores are progressing, we use the newest technology in our spaces so that our designs not only look aesthetically pleasing but can keep up with the high demands of the market today. New technology will bring about new materials and manufacturing techniques which breathe new life into old ways of designing interiors, however the assumption that the future lies in E Shopping and some kind of "Jetsons" modernity is probably a fallacy. The future will take on all kinds of forms, for instance a return to basics and utilitarian sensibilities as well as sustainable practices of design and building.

www.gregnatale.com

1.How can you deal with the clients or their different opinions between your design and client's demand?

Ultimately the client is always right but every so often it is our duty to educate the client and offer suggestions of things that the client might not have considered. Mostly a client will come to me because firstly, they have seen the work I do and like it but ultimately because they trust me.

2.Your ultimate shop project...

YSL Stores, but also anything where I have complete design freedom, where budgets and time and site restrictions don't hinder the final design outcome.

3.You wish you'd designed...

The Gucci Stores

4.Your favorite three shops are...

Tom Ford's New York Store
Dolce and Gabanna's LA Store
Dedece/Minotti Showroom in Sydney

我们看待买卖的态度一直与对待各类设计的态度一致，为我们的客户找到最令人兴奋而奇特的设计方案。在零售过程中，品质尤为重要。零售店的设计主题要做到能时时更新，永远站在潮流的最前端，这极为重要。和所有设计的宗旨一样，创造一个可用的空间，使拉伸设计的局限，把新奇有趣的东西融入你的日常空间。我们利用强烈的色彩调和、大胆的图形、游戏和高品质并存的态度，还有创造出有叫好率的作品的激情实现了这一宗旨，同时我们的成果也给客户带来了无限的快乐。

在对零售店的设计中，我们一直寻求一些方法，创造出不仅仅是为我们的客户量身定制的氛围，还要为使用者考虑更多。随着商店的发展，我们运用了最新的技术，设计出来的作品不仅看着美观，还能跟得上市场的高要求。新技术会带来新的材料和制造工艺，为过时的内饰设计注入新生，那些关于未来只有网上购物和“杰森式”现代化的假设可能成为谬论。未来的购物会以多种方式并存，既有基本实用心态的回归，也有设计和建筑的长期实践。

1. 你如何处理与客户之间的关系，以及他们对你的设计要求提出的不同意见?
客户永远是对的，但我们的责任是启发客户，提出他所没有想到的意见。很多时候，客户来找我，是因为他们看过我设计的作品，喜欢我的作品，所以最后他们总会相信我。

2. 你最满意的商店设计作品是……
圣罗兰店。我对这家店的设计完全自由，预算、时间和场所的限制并没有影响最终的设计成果。

3. 你最想设计的是……
古奇店。

4.你最喜爱的三家店铺是……
汤姆·福特纽约店；
D&G洛杉矶店；
德德西/米洛提在悉尼的展示厅。

GUSTAVO SLOVIK

What catches a customer's attention in a store? What makes one display more inviting than another? What type of layout encourages lingering, longing, and -- most importantly -- buying?

As any dedicated shopper will tell you, a well-designed shop is not merely a commercial establishment, but a leisure experience, an adventure that always comes accompanied by intense doses of emotion and expectation that has to be savored at length. Shops are becoming mini-universes which shoppers can wonder at, be inspired by—and, of course, make their favourite shopping destinations.

People like to feel that they belong to a certain place to be more comfortable and spend more time and money at that particular store; we are creatures of habits and we like to go shopping when we are happy (we deserve it as a price), and when we are sad we go more shopping (we deserve it as a consolation), shopping has become our favourite activity in urban areas; it has become part of our lifestyle

With the increasingly important cultural role being placed upon shops and boutiques in today's competitive retail environment, the need to provide new and innovative shopping experiences has brought about a redefinition of simple consumism. Fashion spaces have taken on the role of the museum, presenting product collections as if they were displaying works of art. With this new emphasis on the role of the retail space, comes the architectural challenge of creating multifunctional, customizable shells that can be renewed to meet the needs of every seasonal collection. Shop interiors in major cities around the world have become increasingly more outlandish and wildly imaginative; the standards are constantly rising, and the parameters of shop design have expanded exponentially. No longer is the retail landscape crafted only for shopping convenience and to reinforce brand identity. It has evolved to reflect the role of shops as a public space where we spend a large part of our leisure time. For this reason, architects and designers are increasingly pushing the boundaries when designing shop interiors. The result: the creation of unique and spectacular spaces within the city; spaces which engage and entertain even the most jaded browsers and add to the city's dynamism.

All kinds of commercial spaces have been remodeled or newly created by adventurous architects and interior designers, each of which is distinguished by more than the products it sells... the interior space, the materials, the colours, the textures, the light, the furniture, the graphic signs, the apparel, etc.

The future shopping

INFORMATION TECHNOLOGIES: For many of us, shopping has become much more than a simple exchange of money for goods and services. It can be a bonding experience, or even a form of self therapy. The past few years have seen the development of new technologies that may completely change the way we shop and how retailers interact with us as their customers. The application of network technologies, the emergence of a new generation of consumers, along with radical innovation in the nature of customer relationships, are likely to make shopping a very different experience in a few years than what we know today.

The Future of Shopping focuses on understanding the evolution of the shopping experience and what changes technologies could bring to enhance the consumer's experience throughout the purchasing process. There are several researchers focusing on creating solutions that provide the consumer with the option to increase control of the retailer in order to enhance their experience and turn latest technologies that are now implemented at the backend of retailing into front end solutions, for the ultimate benefit of both consumers and retailers.

Although there will be tons of new technologies that will make life easier there will always be certain areas where customers will not accept a high-technology store.

MATERIALS: Each age is marked by materials that identify social, cultural and technical aspects of the period. At the beginning of

在商店里，什么最吸引顾客的注意？如何让你的店面更招徕顾客？什么样的布局更令人流连忘返，最重要的是让人更想买东西？

经验丰富的顾客会告诉你，设计精美的商店不仅仅是一个商业机构，还是一个放松的场所，是充满激动和期待的冒险之旅。商店成了微缩的世界，在那里顾客能体验惊喜，鼓舞，当然最后成为他们最喜爱的购物场所。

人们喜欢置身于一个格外舒适的场所，他们愿意在特别的商店里花更多时间和金钱。我们是受习惯驱使的生物，开心时出去购物，我们把它当作奖励，而在伤心时也要购物，它就成了安慰。在城市里，购物是人们最喜爱的活动，它已成为人们生活方式的一部分。

如今的零售行业竞争激烈，日益重要的文化角色被各家商店和精品店置之脑后，人们对创新购物体验的需求为简单的消费主义重新做了界定。大量设计时尚的店面涌现出来，像展示艺术作品一样展示商品。零售店中时尚的盛行，也为建筑带来了挑战。它要求多功能、可定制的房屋骨架，可以时时更新，以满足各季展品的需求。在世界主要城市，商店内部变得越来越古怪，富于疯狂的想象力，标准不断提高，设计界限也以指数方式扩大。零售店的美观形象不再仅仅是为了方便购物和加强品牌意识。它已演变成一个公共场所，人们把大量业余时间都花在那里。基于这个理由，建筑师和设计师在设计店铺内部时把底限推得越来越远。结果是，创造了独特而壮观的城市景观，这些商店鼓励并娱乐了早已厌倦逛街的人士，为城市增添了活力。

各色商业网点都已经被富于冒险精神的建筑师和室内设计师重新诠释，这些店面的内部风格、装饰材料、色彩、质地、光线、家具、手绘图案、外观等本身特性比它们所出售的商品还要吸引人。

未来时代购物

信息技术：对我们许多人而言，购物已不仅仅是简单的金钱与商品、服务之间的交换。它是一项综合性的体验，甚至是一种自我治疗。在过去的几年中，新技术的发展也许彻底改变了我们的购物方式，以及零售商与顾客的沟通方式。网络技术的应用，新一代消费者的产生，以及对顾客关系本质的激进创新，大概在未来几年，购物的体验将会和今天大不相同。

未来购物的重点在于了解购物体验的演变和什么样的技术变革能在采购过程中提高消费者的经验。这需要研究人员为消费者提供有选择的控制零售商的途径，以提高他们的经验。还要改变最新技术，使现在实施的后端零售转化为在前端解决，最终达到消费者和零售商的双赢。

虽然大量新技术让生活变得更轻松，但在有些特定领域中，总会有消费者无法接受的高科技商店。

选材：各个年代都有标志性的材料，显示了当时的社会、文化和技术。20世纪初期，玻璃钢的引进从根本上改变了设计方法。包豪斯宣言引导人们使用不同的物质，将物质性的研究推广到各个领域。

现在，原材料的可用性大大增加，带来了信息的发展。信息爆炸造成了应对物质文化的设计师和拥有较高技术专业知识的生产者之间的对话障碍。而由于规模和经济原因，材料技术的进步也比设计实践快了很多。

随着设计师和生产者之间的鸿沟渐渐缩短，会有更多设计师利用更广泛的原材料和更具创意的方案，创作出新的作品。

在SLVK，我们经常利用已知的原材料，以不同以往的方式把这些原材料应用于生产过程中，展现出新的表现力。我们用材料的不同使用方法来满足客户的需求，这在一定程度上是一种原始的创新。

在本书中，我们会看到很多设计精美的店铺，其中绝大多数店铺坐落于大都市，有些依傍在古典建筑之上，但都以当代的设计方法加以解决，有些使用了原始的装修方法，利用了大量装饰材料和色彩，还有些店面更多使用了灯光和家具，或引入了不一样的织物和地板。我可以对每个项目进行分析，找出其中有趣的特点，但这是读者应该做的事……

the 20th century the introduction of glass and steel radically changed the approaches of design. The Bauhaus Manifesto induced experimentation on and with different matter, introducing the study of materiality to several fields.

Today the increase in the availability of material has given rise to increase information. This excess of information has created a language barrier between designers who deal with selected material culture and producers who have highly technical, specialised knowledge; also the advances in material technologies are going much faster than design experimentation for reasons of scale and economic concerns.

As this gap between designer and producers is reduced there will be more designers using a wider range of materials resulting in newer projects with more creative resolutions.
At SLVK we always like to experiment using known raw materials that are already part of the production process on different ways than the usual, achieving new forms of expression and different uses that permit us to resolve our clients' needs in an original manner with a certain degree of innovation.

In this book we are going to see very good designed stores most inserted in an urban context, some in classical buildings facades but resolved according a contemporary design approach, some with original approaches made with a great variety of materials and colours, others playing more with the lighting design and furniture or introducing different textiles and flooring. I could analyse each of the projects and find lots of interesting features in each of them but that will be the reader's job to discover …

Interview
1)How can you deal with the clients or their different opinions between your design and client's demand?
There are two simple rules you must follow in the client-designer relationship to achieve the best results:
First, understand your client and their necessities and you have to take as much time as necessary to do that and after that your design should be a response to those necessities.
Second, use all the means at your hand to explain your client the design scheme, its reasons and expectations, where it comes from, why you have chose the materials, what are you expecting to achieve with the layout, lighting and colours in your design… everything should be done with a reason. You should include your client as a part of the design team, let him be part of the decisions and specifications, listen and understand him and then you´ll be able to answer all his question and doubts in a favorable manner.

2)Your ultimate shop project...
I liked very much to design "SPECIA" gourmet store in Mexico City it was a big challenge…
The original state of the store had a timid outward presence due to its generic image. One of the issues to consider was to look for the potential customer winning. Thus, it is proposed a clear and marked access, supported by an original and seducing image.
The limited budget demanded a greater emphasis on creativity promotion. The proposal for "Specia" commercial store avoided at the most the expense on non essential elements and stresses efforts on elements allowing for an eventual occupation: furniture, shelves, and image. Through a solid and identifiable image it is looked ….
THE IMAGE CONCEPT was based on the new kitchen, created by the major change carried out by the French "Nouvelle Cuisine" at the end of the 1960s and represents a way to access gastronomy breaking taboos introduced by traditional kitchen. This culinary revolution extended later to all European countries, where in opposition to traditional kitchen, there was a more stressed interest to care for meals presentation and to resort to a wide variety of spices to scent foods. All that was useful to design the store image, a precedent to maintain an open attitude based on the search for new products and to create a meal considering it was much more than a simple ingredient mixture: the goal was always to surprise with anything different. At "Specia" the raw material was an element; an autonomous ingredient that

did never use its paper within an artisan process.
We decided to use cardboard tube (used mainly to contain the concrete in rounded columns) as the main material because of its price (very cheap) and its image that permitted all the grains, species and cans to be the main performers of the store because of their colours and textures.

3)You wish you'd designed...
I have always wanted to design a museum store, I like to walk around those stores, they always have something interesting, books, design items, posters and tons of useless gadgets and sometimes you find something really useful.... I also like alessi type items very much so it would be nice to a store where we can use all our imagination and creativity for the display.

4)Your favourite three shops are...
I like the stores I'm mentioning because they do share the same ideas as the brands they represent or contain in a natural way. I like Apple Store Fifth Ave by Bohlin, Cywinski, Jackson, because of the great glass façade and lighting which really reflects all the apple statements of design vanguard and simplicity. The Prada Flagship Store in Soho by Rem Koolhaas for the innovative interiors so simple and yet excellent (the auditorium is great) and Vinçon Barcelona a store you have to visit with lots of time to enjoy the cool items they sell.

1. 你会如何处理与客户的关系，以及他们对你的设计要求提出的不同意见?
客户和设计师的关系，有两个简单的规则你必须要遵循，这样才能达到最佳效果：
1)了解你的客户和他们的需求，你必须花相当长的时间来做这件事，针对那些需求来设计。
2)利用现有的一切手段向你的客户解释设计方案，其理由和预期效果，包括方案从何而来，你为什么选择那些材料，在设计方案中你期望达到什么样的布局、灯光和色彩……所有一切都要有原因。
你应该把客户当成设计团队中的一员，让他参与决策和细节，倾听他，理解他，接下来你要以合适的态度解答他所有的疑虑。

2.你最满意的店面设计作品是……
我最喜欢对墨西哥城“SPECIA”餐馆的设计，那是个大挑战……
那家餐馆原来的门脸非常小。其中一个需要考虑的问题是吸引潜在顾客。因此，我们为它设计了一个鲜明显眼的店门，上面有原始而性感的图案。
由于预算有限，我们更加强调创造力的发挥。“SPECIA”的设计计划是避免大量非必需品的支出，增加空间占有物的使用，如家具、架子、挂画等。通过可识别的实体形象打造这家店面……
形象的概念以新厨房为基础，起源于20世纪60年代末法国的“新潮烹调”所引发的重大改变，代表了以烹饪法打破传统厨房禁忌的方法。随后，这次烹调革命席卷了所有欧洲国家，人们都在反对传统厨房，更加注重食品的外观和使用各种香料。所有这些都对设计店铺形象有所帮助，有一个开放观念的先例，勇于寻找新的产品，创造了烹调理念，烹调并不仅仅是简单地把原料混在一起：目标始终是因不一样的事物而惊喜。“SPECIA”的素材就是一个设计元素，是原始的，从未经过任何人工雕凿。
我们决定用纸板管（主要用于固定混凝土柱子）为主要材料，因为它的价格非常便宜，造型也适宜。粮食、香料和罐头的色彩丰富，纹理也很漂亮，我们也让它们成为这家店的主角。

3.你最想设计的是……
我一直都想设计一个博物馆商店，我平时就喜欢逛这样的店，它们总卖些有趣的东西，书籍、设计项目、海报、一堆没用的小玩意，有时你还能找到些真正有用的东西……我还很喜欢阿雷希风格的东西，这样的商店真的很棒，我们能发挥全部想象力和创造力来展示商品。

4.你最喜爱的三个商店是……
我喜欢以下商店，因为它们有的展示了与其品牌相同的理念，有的设计得十分自然。我喜欢波林·赛温斯基·杰克森设计的第五街苹果店，巨大的玻璃幕墙和灯光真正反映了苹果的设计理念——前卫和简洁。雷姆·库哈斯设计的索霍区普拉达旗舰店，创新的内部设计如此简单，又如此出色（礼堂的设计尤为出众）。还有巴塞罗那的文康店，你一定要多去几次，他们卖的那些东西特别棒。

Guy Zucker—Z-A Studio

Every consumer of fashion is aware of the fact that he/she is paying for design rather than material. Furthermore, the pace at which an item of clothing becomes obsolete is dictated by the change of fashion trends and seasons and not by the quality of the item. Therefore the capital invested in design exceeds the funds invested in quality. In architecture the balance of investments is reversed, the cost of materials far exceeds the investment in design.

What we can learn from mega flagship stores like Prada or Apple in Soho, is that even if you set up a store with the intention of it being more a display case and part of an advertising campaign, once people are inside they want to participate in the experience. It's like buying a souvenir when you travel to a place you are excited about. The idea of blending shopping and art, high and low culture was already well discussed in the last decade. I think no one will argue that we all spend time shopping not out of need, that when we want to see design we can go to MoMA or we can go to NoLIta. But there is something else, the internal logic of fashion design, its economics, its pace; there is something there that is extremely intriguing for other design fields to learn from, to work with. Fashion design has managed to base its entire economy on the idea of experience not need, a notion that is magnified by its fast pace of seasonal changes. This structure will be appropriated into other design fields for two reasons, firstly its unquestionable marketing structure and secondly as we all gradually realie that our environments can now be transformable with growing ease.

Fashion design is a creative field that has managed to completely merge with contemporary life. The entire industry is based on global reality and constant transformation. Architecture has to confront the same reality of ever-changing programs, budgets and building uses. If we want to learn from fashion design, we have to change our expectations from architecture. If we don't expect all architectural products to become monuments, if we can invest less in 'high-end' materials, we might be able to give more importance to the manipulation of material and the quality of design rather than the cost of matter, to the designer rather than the contractor.

1.How do you deal with the clients or their different opinions between your design and client's demand?
Maybe the best answer to this question would be by explaining the design process of the Delicatessen clothing store. The brief that was given to me was pretty basic: display elements, fitting room, counter, storefront and design studio. The important part of the brief was located only when we started the process. I think that at the end it is always up to the architect to conceptualise the program and find the hidden narrative behind the basic needs of the client. Before we started the design I tried to establish an understanding of what constitutes the commercial markets and design goals of a boutique as opposed to a well-known brand (considering the fact that prices are similar). Through a conversation with the client (the fashion designer) we realised that the 'smallness' is actually an advantage and that clients were looking for that close connection with the designer. This notion led us to consider the design of the studio space as part of the store and to emphasise that connection. The organisational scheme of the three bands that roll up like conveyor belts out of the studio and into the store was born out of this notion.

2.Your ultimate shop project...
My ultimate shop project would actually be something that I have already started to work and would like to see it made into reality. Earlier this year I have been hired by a New York developer to design a vision for recycling the American strip mall. One of the schemes is for using the existing generic structure as a shell that hosts temporary stores. Current trends in online shopping push physical shopping even further into the realm of entertainment. Consumers are willing to spend much more on experience than on actual goods and at the same time the stimulation threshold is elevated. The design takes advantage of the numerous

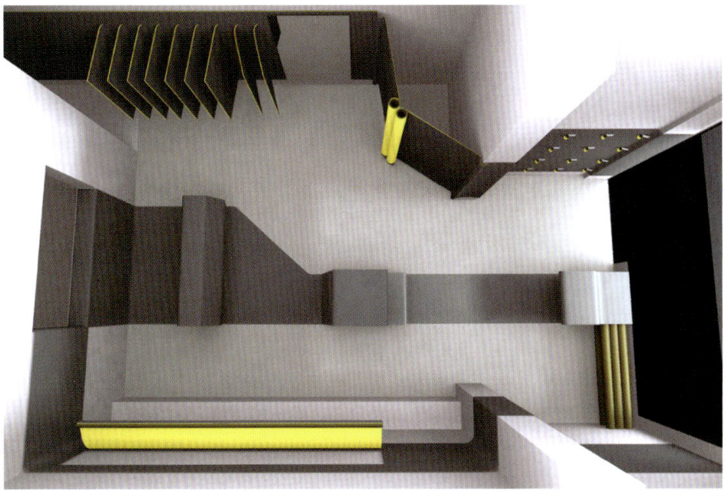

dead strip malls scattered throughout the American landscape and conceives of them as a network of hosting shells. A secondary network of pop-up stores circulates between the stationary hosting mall structures. The mall turns into a cyclical event venue where the stores themselves, rather than the merchandise, fill it with new content.

3.You wish you'd designed...
I would like to design the Wall-Mart stores worldwide.

4.Your favourite three shops are...
My favourite three retail projects to date are the three projects that appear in this book for the simple reason that they are the only commercial projects that I have designed and built so far.

每一个买时装的人都知道，事实上他/她买的是设计而不是面料。此外，时装过时是由于时尚潮流和季节的变化，并不是品质的问题。因此，资本对设计的投资超过了在质量上的投资。而在建筑行业，投资平衡是相反的，在材料上的花费要远远高于设计费用。

我们可以借鉴大型旗舰店，如索霍区的普拉达或苹果店。假如你开了个店，加大商品展示和广告的力度，人们进店来就是想参与体验。这就像旅游的时候买纪念品，会让人很兴奋。

购物与艺术、高级文化与低级文化相融合的理念在过去的几十年中早已得到充分论证。我们花费大量时间所购买的并不是没用的东西，我想没人会对这提出异议。想欣赏设计作品的话，我们会去纽约现代艺术博物馆或去诺莉塔。但还有个问题，时装设计、行业经济和发展的内部逻辑是学习借鉴其他领域的设计方案。时装设计业已经设法把自己的整个经济重心建立在经验无用的理论之上。随着时节的变化，该行业的发展步伐加快，这一理论也被夸大。这种结构也适用于其他设计领域。有两个理由，首先是其不可置疑的营销结构；其次，人们逐渐意识到环境变得越来越舒适，环境也是可以改变的。

时装设计是一个创造性的领域，它已完全融入当代生活之中。整个行业建立在总体现实与不断变化之上。建筑业也面临着同样的现实问题——不断更新、预算和建筑用途。我们要向时装设计业学习，必须改变对建筑的期望。如果我们不再期望所有建筑都成为不朽的丰碑，如果能减少高端原材料的投入，我们就会更注重原材料和质量的设计应用，而不是成本问题；更注重设计师，而不是承包商。

1. 你如何处理与客户之间的关系，以及他们对你的设计要求提出的不同意见?
也许对这个问题最好的回答是对“成衣”服装店的设计过程加以解释。总的来说，基本的方面有：展示区、试衣间、柜台、店面和设计工作室。设计工作开始时，最重要的是定位。我想，最后总是建筑师把项目概念化，找出客户基本需求以外的隐藏需求。设计开始之前，我试着了解市场的构成和精品店要与名牌竞争的设计目标（考虑价格相近的因素）。通过与客户（时装设计师）的谈话，我们发现“小”其实是个优势，这个客户想跟设计师近距离交流。基于这一想法，我们打算把工作室设计成商店的一部分，以强调关联性。由此产生的三组设计计划立即成型打包，就像传送带一样，从工作室输送到店里。

2. 你最满意的商店设计作品是……
我已经开始着手最终的商店项目，并会亲眼看着它成为现实。今年早些时候，纽约的一个开发商聘请我设计一个可循环再利用的美式沿途商业区。其中一个计划是把现有的通用结构当作载体，来承载那些临时商店。目前，网上购物的趋势促使实际购物向娱乐性方向发展。跟商品相比，消费者宁愿把更多的钱花在获得经验上，同时刺激消费的起点也有所提升。设计利用了美国本土大量已经倒闭的商业区，把它们当作提供载体的网络。新兴商店的次级网络在不变的商业区建筑之间循环。商业区变成了周期性的活动场所，商店本身，而不是商品，则会为它带来新的内容。

3. 你最想设计的是……
我想设计全球的沃尔玛店。

4.你最喜爱的三家店铺是……
我最喜爱的三个商店项目都出现在本书之中，理由很简单，它们是我至今设计和建造过的唯一的商业项目。

Hartmut Raiser / Kristina Lopes

We have been planning stores, shop in shop and outlets for years, but day after day we questions our work and search for new solutions. We try not to repeat our work, since every environment is different. Such as fashion undergoes changes permanently, without denying its roots, architecture is latently challenged by the demands on novelties.

Clients call for more atmosphere or a unique atmosphere and more individual presentation. More than ever shopping pictures a great experience. To see and be seen with small or big shopping bags is still important. (Free delivery is just interesting for food.)

As architects it is a pleasure to deal with regional circumstances, but nowadays limits are very tight. The brand's CI is the most important criteria for the positioning on the market, for this reason shop design has to proceed gently, since customers want to recognise their favourite brand. Variety in appearance of special brand stores or shop in shops are counterproductive.
But for all that shop design has to change and respond to the situation. This is a balancing act and these concepts need a strong design which can be modified gently.

Cool Shop shows definitely the trends of shop design and gains insight into high-value international developements of various industries. Less "old brands " but mainly young and new brands are represented. Fresh design with a tendency towards extroversion and nonchalance, a wild time in a young milennium. Shouldn' t we let these chances and sources of unlimited inspiration have its effect and just do crazy things from now on?

1.How do you deal with the clients or their different opinions between your design and client's demand?
A good concept grews from listening to the clients' demand and our dealing with the subject in a repeating process. A common aim for both has to be clearly in mind. Of course we respect different opinions, but we try to extract statements and try to get to the bottom of their intension. This way we get a picture of both the client and his challenge.

2.Your ultimate shop project...
The Store for Hugo Boss Orange in Xintiandi/Shanghai was our ultimate project in the last years. It was a new development and we used unusually poor materials for a conceptual and architectural solution.

3.You wish you'd designed...
I wish I had designed the interiors for James Bond after Ken Adams

我们设计店铺、店中店和商行已经有好几年的时间了，我们仍然经常对自己的作品提出疑问，并寻求新的解决方案。我们不想在作品中出现重复，因为每个环境都有其不同之处。虽然时尚在不停变化，但始终离不开其根本，为满足人们对新奇事物的要求，建筑就成了潜在的挑战。

客户要求更多独特的气氛和更多体现个性的东西。如今，购物体现了一种伟大的经验。看见别人和被别人看见手中提着大大小小的购物袋仍然十分重要。（人们只对食品的免费送货服务感兴趣）

作为设计师，能为一个大环境做设计很高兴，但如今各种各样的拘束也比比皆是。一个品牌的企业形象识别是做市场定位的最重要标准，客户也想要认知他们喜爱的品牌，因此，做商店设计时都要考虑周全。但是，所有的商店设计都改变了这一局面。这是一种平衡的举措，需要有一个强大的设计理念来慢慢改变这些概念。

“酷店”展现了商店设计的潮流趋势，获得了洞察各行业高价值国际发展的眼光。酷店主要以年轻的新品牌为代表，老牌子很少。新鲜的设计风格趋于张扬和冷漠，在新千年里掀起野性风潮。我们是不是应该给这些机会和无限灵感以用武之地，从现在开始做疯狂的事?

1. 你如何处理与客户之间的关系，以及他们对你的设计要求提出的不同意见?
好点子来源于我们倾听客户需求和处理项目的重复过程。设计者的脑海中对二者的共同目标十分清楚。当然，我们尊重不同意见，但我们会尝试阐述我们的看法，并试着寻求解决办法。

2. 你最满意的商店设计作品是……
上海新天地的雨果·波士橙色店是我们去年的顶级之作。那是个创新的作品，我们用平时不常见却很便宜的材料来解决概念和建筑上的问题。

3. 你最想设计的是……
我想做肯亚当斯后的007内部设计。

Paul Cha

Due to the rise of Internet and multinational commerce, the shop design had undergone a major evolution in the last decade. The traditional department store such as Barney's, used to serve as an communicative advertising medium where shoppers can be educated and seduced by varieties of options. Shoppers today will do their research on line first before arriving at shops with definitive preferences, and their shops are designed to be receptive to these preconceived preferences.

In short, the traditional department store where massive amount of goods are displayed in corporate setting, had been replaced by smaller, more focused specialty shops where goods are selected and displayed by specific aesthetic considerations. The design of such specialty shops can be divided in two dramatic opposite approaches. The first approach treats the shop as an extension of the designer's aesthetic outlook. Just as the particular cut and finish of a designer's clothing, the shop's layout, material, finish & detailing compliment the same design vision. The Lanvin store as illustrated in this book, and Jil Sander store exemplify this self-contained approach- the same design strategy is carried throughout the world. The second approach transports the "feel" of local neighborhood surrounding into the store itself, so the store recesses into a neutral framework while the merchandise's uniqueness progresses into the foreground. Two interesting examples are Colette in Paris, where the store captures the surrounding Classical Parisian chic, while Opening Ceremony in New York, where the neighborhood mood is distinctively Bohemian Soho hip. These two stores are vernacular and regional in character, and the stores' authentic atmospheres are hard to transplant anywhere else in the world.

Just as the store design had evolved, shopper's attitude toward the merchandise had evolved as well. The recent phenomenon of fashion items' elevation into objects of fetish desires, as shown in Karl Lagerfeld's Chanel designs cultishly displayed in Zaha Hadid's Chanel store, parallel to the art world's inverse treatment of everyday items into objects of worship, as shown in Jeff Koon's Balloon Dog, 1994-2000. Hence shops has become temples, where desirable merchandise are loving displayed and admired, while the subtle sensual pleasure of ownership will persuade the shopper to complete the commercial transaction at the checkout counter. In essence, the store and its merchandise can be a projection of the shopper's personal identity, just as the shopper's home and its contents reflect the owners' taste. To strengthen such religious personal projection, the store's clever architectural layout of Entry, Display, Change Room/Storage, and Checkout is of paramount importance, while the architectural detailing, such as material and finish, as well as colouring and lighting, lavishly illustrated and photographed in this book, will compliment and reinforce the unique sensuality of products being sold.

Ultimately, a successful store design is not only a three dimensional manifestation of a brand's aesthetic identity, but it also elevates the shopper's private desire for object ownership into an urban communal interaction where shoppers gather, interact, touch and purchase. Shopping had evolved from the purchase of everyday necessities into a collective expression of individual personal preferences- transcend any social, economic, ethnic and age barriers. An elegant store design reinforce the dialogue between the shopper and the merchandise- that of material longing, sensual seduction, and personal fulfillment.

1,How can you deal with the clients or their different opinions between your design and client's demand?
The shop is essentially a physical framework encasing commercial display, emotional persuasion, and financial transaction. Where the differences arise between the Client and Designer are the creative means to achieve that end. The Designer needs to initiate the design process with a strong aesthetic vision in harmony with the Client's store identity and financial boundaries. The Client's

aesthetic doubts and financial challenges during the design process can be channeled into clarifying and editing the Designer's original intention, hence strengthening the store's final design.

2,Your ultimate shop project...
To design the ultimate "Lifestyle" store in Shanghai, China. The store would encompass fashion, art, bookstore, antique, house ware, cafe and restaurant. Similar to Carla Sozzani's 10 Corso Como in Milan, Italy.

3,You wish you'd designed...
Prada in Soho, New York. A 2001 zeitgeist combination of Soho's loft-like spatial sensibility, Rem Koolhaas(OMA)'s architectural outlook, and Miuccia Prada's cutting edge fashion creations.

由于互联网和跨国商业的发展，在过去的十年中，商店设计经历了重大的变革。巴尼等传统百货公司，过去就是交际广告媒介，在那里，购物者会受到启发，并被各种各样的选择所引诱。今天的购物者如果有了购买目标，在进商店之前，他们会先在网上搜索一番，为他们打造的商店要设计成可以接受这些先入为主的偏好才行。

很快地，大量商品共同展示的传统百货公司被规模更小、更有针对性的专业商店所取代。它们的商品经过精挑细选，展示时也会考虑美观因素。设计这样的专业商店有两种截然相反的方法。第一，是把商店当成设计师审美观的延伸。时装设计师对设计的服装精雕细琢，同样的店内布局、选材、完工，每一个细节的设计理念完全相同。书中介绍的浪凡店和吉尔・桑达店就是这种自我包含方法的例证，同样的设计策略在世界各地都有应用。第二，是把“感觉”放到商店所在的邻近地区之中，模糊商店的轮廓，商品的独特性则变得尤为突出。有两个有趣的例子。巴黎的科莱特时尚商店有着周围古典巴黎时尚的气息，而纽约的开幕庆典，周围就都是特立独行的波希米亚嬉皮风格。这两家店都有着浓郁的地方特色，店内的氛围是世界其他地方都学不来的。

随着商店设计的演变，购物者对商品的态度也发生了变化。最近的时尚提升为恋物欲，卡尔拉格斐为夏奈尔设计的作品被陈列在扎哈・哈迪德的夏奈尔商店中顶礼膜拜。1994年至2000年，杰夫・古店里的气球狗则显示了艺术世界对物质崇拜的相反待遇。商店变成了圣堂，人们热衷于展示和艳羡梦寐以求的商品，而拥有它这一微妙的感官愉悦促使购物者在结账柜台完成了交易。本质上，商店和商品能折射出购物者的个人特质，就像一个人的家和家中物品能反映出他的品味。为了加强这种个人特性的折射，商店对入口、展示区、试衣间/寄存室、付款台的精心布局就显得无比重要。建筑设计必须详细化，从原料到完工，包括色彩、灯光，配合和加强了出售商品独特的诱惑力，这在本书中有大量的说明和照片。

最后，一个成功的商店设计不仅是一个品牌美丽特质的三维展示，它还提升了购物者把其据为己有的欲望。城市里消费者云集，他们相互交流，触摸商品，然后购买。购物从采购每天的必需品，演变为个人喜好的综合表现，它超越了社区、经济、种族和年代的界限。一个优雅的商店设计能增进购物者与商品之间的对话，促进物质渴望、感官诱惑和个人满足。

1.你会如何处理与客户之间的关系，以及他们对你的设计要求提出的不同意见?
店铺实际上是一个框架，里面有展示的商品、情感安抚和金钱交易。客户和设计师最易产生分歧的是用什么样的创意手法来达到效果。设计过程开始时，设计师要运用独特的审美眼光，找出与店铺的特性和预算最为和谐的方案。设计过程中，客户对于设计效果和经济支出提出质疑，设计师研究解决方案并对最初的设计加以修改，以使店铺最终的设计效果更加完美。

2.你最满意的商店设计作品是……
设计中国上海的“生活方式”店。那家店包括时装店、艺术品店、书店、古董店、家居用品店、咖啡厅和餐馆。跟意大利米兰卡拉・索珊尼的科摩差不多。

3.你最想设计的是……
纽约索霍区的普拉达店。2001年时代精神的结合包括索霍区的阁楼式建筑，雷姆・库哈斯（大都会建筑事务所）设计的建筑景观和缪西亚・普拉达设计的前卫时装。

Alannah Hill Store
Brisbane, Australia

阿拉纳 希尔商店
澳大利亚，布里斯本

The designer believes that the possibilities of design are endless and constantly shifting. Fashion may inform the design ideas, but there are limitless directions for interpretation and definition. Strong planning and an understanding of the pragmatics of design achieve "recognisable and unique, sophisticated and contemporary" balance, while choice of materials and finishes outstay "trend" and "fashion".

Alannah Hill is a sexy brand which remarkable for the colour, pretty and dazzling. The key design concept here is to evoke the feeling of an elegant, sensual boudoir. The scale of the spaces,the finishes,lighting and décor all work together to realise this effect. Into these rooms familiar, domestic scale pieces have been carefully placed. Chandeliers give a feeling of opulence, while fringed lampshades offer an intimate touch. An ambience of old world charm is created by the carefully selected components, from the ornately moulded tables and chairs, to the elaborate light fittings, and sumptuous fabrics.

设计师深信设计的可能性是无止境并不停变幻的。时尚可以显示出设计理念，但也有无数的方式来阐述和定义。完善的计划和对设计语言的领悟能够使设计实现“大众认可与个性独特，古典与现代”的平衡，同时材料的选择和修饰使潮流和时尚更加持久。

Alannah Hill是一个以色彩与美艳著称的性感品牌。Alannah Hill商店的主要设计理念就是营造一个优雅的闺房。空间的比例、光泽、照明以及装饰共同达到了这样的效果。每个空间摆放的物品都经过精心选择，并传递着熟悉亲切的感觉；枝形吊灯让人感觉优雅、富贵；流苏灯罩散发亲密感。华丽的桌椅、精致的灯光设备以及高雅的布艺尽显古色古香的魅力。

PROJECT
Alannah Hill Store
DESIGNER
Hecker Phelan
Guthrie P/L
COMPLETION
2006
PHOTOGRAPHER
Diephotodesigner.de

项目
阿拉纳 希尔商店
设计师
Hecker Phelan
Guthrie P/L
完工时间
2006
摄影师
Diephotodesigner.de

EXIT

EXIT

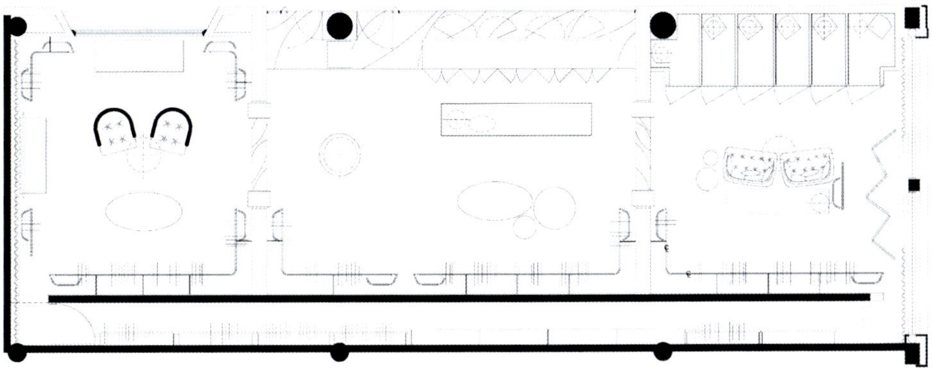

BIASA BOUTIQUE
Jakarta, Indonesia

BIASA服饰精品店
印度尼西亚，雅加达

Realized in 2007, in Jakarta, in the Kemang district.The main expressive point is the reinforced concrete staircase in the heart of the boutique. These elements are profiles with a rectangular section folding up following a G shape. They have been realized on site in reinforced concrete in different heights following the height of the steps. The upper part of the G profile is the step of the staircase, while the lower one becomes the exhibition of the shop.

It's not a traditional staircase anymore . It follows this idea of mixing two different functions, the vertical connection and exhibition. Moreover this staircase is stand alone by the structural point of view , because it is not connected to the wall, but each G element sustains itself. The strenght of the staircase is doubled by a bronzed mirror wall, 23 meters long for a total of 125 square meters, which amplified the space and contains useful hidden wardrobes for the shop.

The Art gallery space is characterized by an open space with a mixing of materials like clear concrete, wood and terpal, a plastic and cheap material used in rice fields by the Indonesian farmers here applied in volumes hanged on ceiling to hide the main lighting of the staircase to these floors.

别萨展厅位于雅加达的克芒区，完成于2007年。最引人注目的是精品店中心坚固的混凝土楼梯。从侧面看，就像一个个长方形混凝土块沿着G字形折叠起来，随着踏步的高度起伏。G形上半部是楼梯踏步，下面则是店内的展示空间。

楼梯跳出传统的模式。它将两个不同的功能融合在一起，即纵向连接和展示。此外，从外观上看，楼梯结构完全独立，它没有连接到墙上，每个G形元素都扮演着自己的角色。铜色的镜子墙面使楼梯的强度增加了一倍。墙面长达23米，有125平方米，不但起到扩展空间的作用，其中还设有隐藏式衣柜。

画廊的特色是开放式空间和各色材料的混搭，如透明凝胶、木材和“特普”——一种廉价的塑料，印度尼西亚的农民把它们用在稻田里，在这里，它们被卷了起来，挂在天花板上，以隐藏楼梯的主要照明设备。

PROJECT
BIASA BOUTIQUE
DESIGNER
Giovanni D' Ambrosio
COMPLETION
2007
PHOTOGRAPHER
Graft

项目
BIASA服饰精品店
设计师
Giovanni D' Ambrosio
完工时间
2007
摄影师
Graft

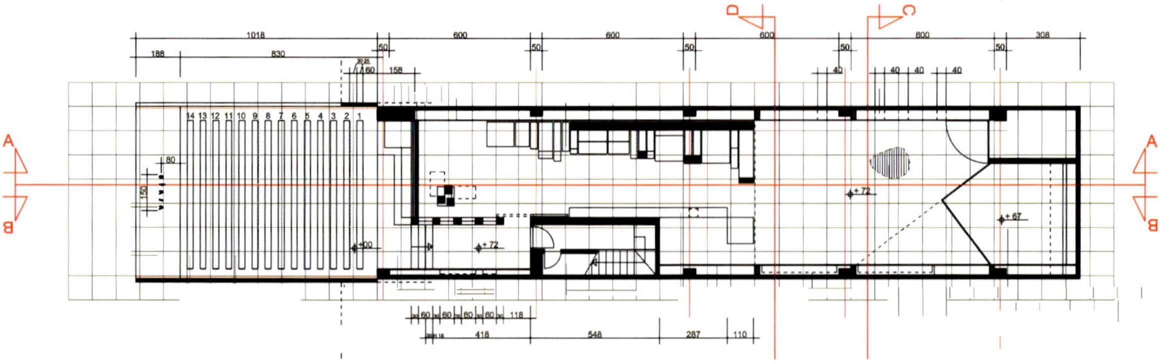

1018
600
600
600
600
308
188
830
50
158
40
14 13 12 11 10 9 8 7 6 5 4 3 2 1
80
150
+00
+72
+72
+67
A
B
C
D
118
416
548
287
110

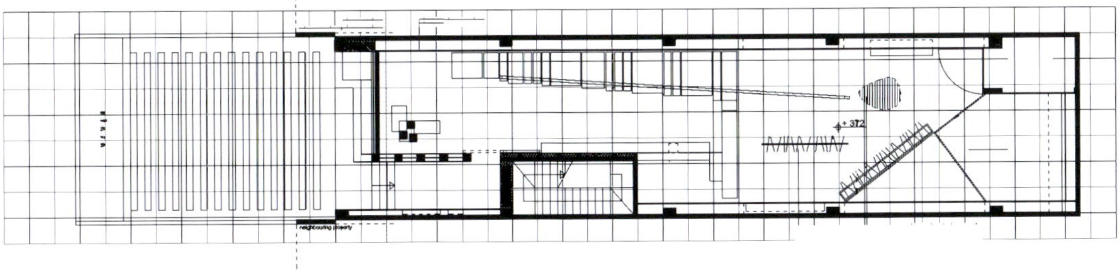
+ 372
neighbouring property

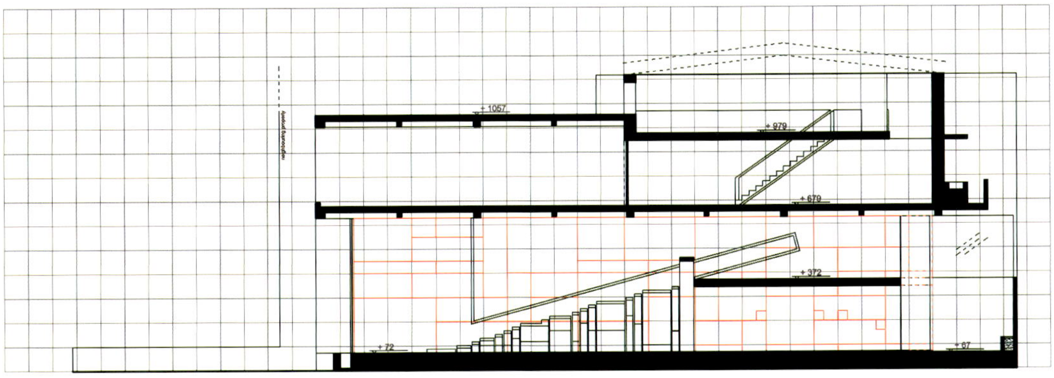
+ 1057
+ 979
+ 679
+ 372
+ 72
+ 67
neighbouring property

BARCLAYS
TSG
restaurant

Boutique Shu Uemura

Paris, France

植村秀精品店

法国，巴黎

The design style must be in tune with "the drive to direct people's behaviours while respecting their desires".

Shu Uemura blends the best elements of "Nature, Science and Art" to create contemporary make-up, innovative skincare and high quality professional tools that fuse simplicity with elegance.

The boutique's main characteristic is its ceiling made of white stratum following the whole depth of the boutique. These successive stratum, layered from floor to ceiling match the various uses of space: make-up countertop, products linear display, sales counter, a basin and its plumbing fixtures and also concealed technical elements. The opposite wall features a shiny black lacquered surface, working as a mirror to the white stratum. This straight black wall holds retro- lit shelves containing and outlining the wide chromatic shades of Shu Uemura's products.

设计的风格需要和“引导人们的行为和尊重人们的愿望”相搭调。

植村秀将“自然，科学及艺术”这些最好的元素混合，创造出当代化妆、肌肤护理新革命以及高质量专业工具，使简洁与高雅实现了完美融合。

这家流行服装店的主要特点是，这个店面内的深度式设计与屋顶上的白色涂层的鲜明对比。这种从地面到屋顶层层排列的白色涂层配合了对空间不同利用：定做的工作台、产品的并排展示、销售柜台、一个洗脸水盆和它的管子附件以及其他隐蔽的技术性设备。笔直的黑色墙壁映衬出产品的特色。

PROJECT
Boutique Shu Uemura
DESIGNER
christophepillet
COMPLETION
2007
PHOTOGRAPHER
Patricia Parienjab

项目
植村秀精品店
设计师
christophepillet
完工时间
2007
摄影师
Patricia Parienjab

tokyo lash bar

BARCLAYS

tokyo lash bar

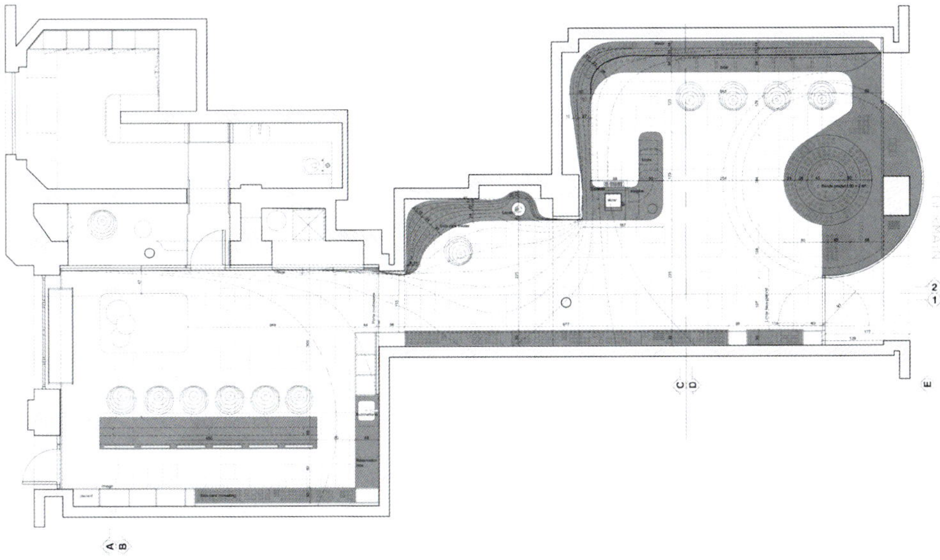

alechinsky peter et pierre
Emilio Trad
THINGS I HAVE LEARNED IN MY LIFE
Annette Messager
FLYER SOZIOTOPE

Bozar Shop

Brussels, Belgium

伯萨书店

比利时，布鲁塞尔

From a spatial point of view, designers kept visible the beautiful Horta's structure. In the same way, they kept visible the former divisions between shops and some stairs and use some of them to divide the different sections of the new space, some stairs are still in use, some other are just traces of the past.

The lighting system is also based on the initial structure (the lights are in the middle of the rectangles formed by the beams) but designers placed the fixtures where needed and connected them to this central light with loose cables. The furniture, conceived as a mix between prefab customized elements, brutalist pieces (materials, structure, construction) and more delicate, designed objects (reminding Horta's time craftsmanship) is a very important aspect of the project. All these elements are designed as small pieces of architecture defining the space around them. The use of different wood species apparently randomly assembled, of different colors arbitrary chosen from a painting by David, produce a feeling of a strong unity as well as a variety of color and texture.

在空间上，设计师采用了明显的奥尔塔式结构。他们保留了部分店面及楼梯，利用它们来分割空间。有些楼梯仍在使用，而有些则成为对过去的怀念。

照明系统是在原有的结构上建立起来的，灯被安装在天花板的中央。同时，设计师在需要的地方加设了灯具，并用松散的光缆将它们与中央的灯具连接起来。定制的家具在材料、结构和建筑方面隐约地显露出野兽派风格，精美绝伦，展示了奥尔塔时期的工艺，是十分重要的设计。这些小元素使整个空间变得别具一格。店内采用不同品种的木料，配上大卫创作的各种颜色随意涂鸦的绘画，创造出丰富多彩的颜色和纹理，建立起强烈的统一感。

PROJECT
Bozar Shop
DESIGNER
Pablo Lhoas
Pierre Lhoas
COMPLETION
2007
PHOTOGRAPHER
Lydie Nesvadba

项目
伯萨书店
设计师
帕伯罗·洛哈斯
皮埃尔·洛哈斯
完工时间
2007
摄影师
莱迪·涅斯瓦巴

BALKAN BLUES

BROOKS
VISCONTI

BOOKS
PAY
BOOKS

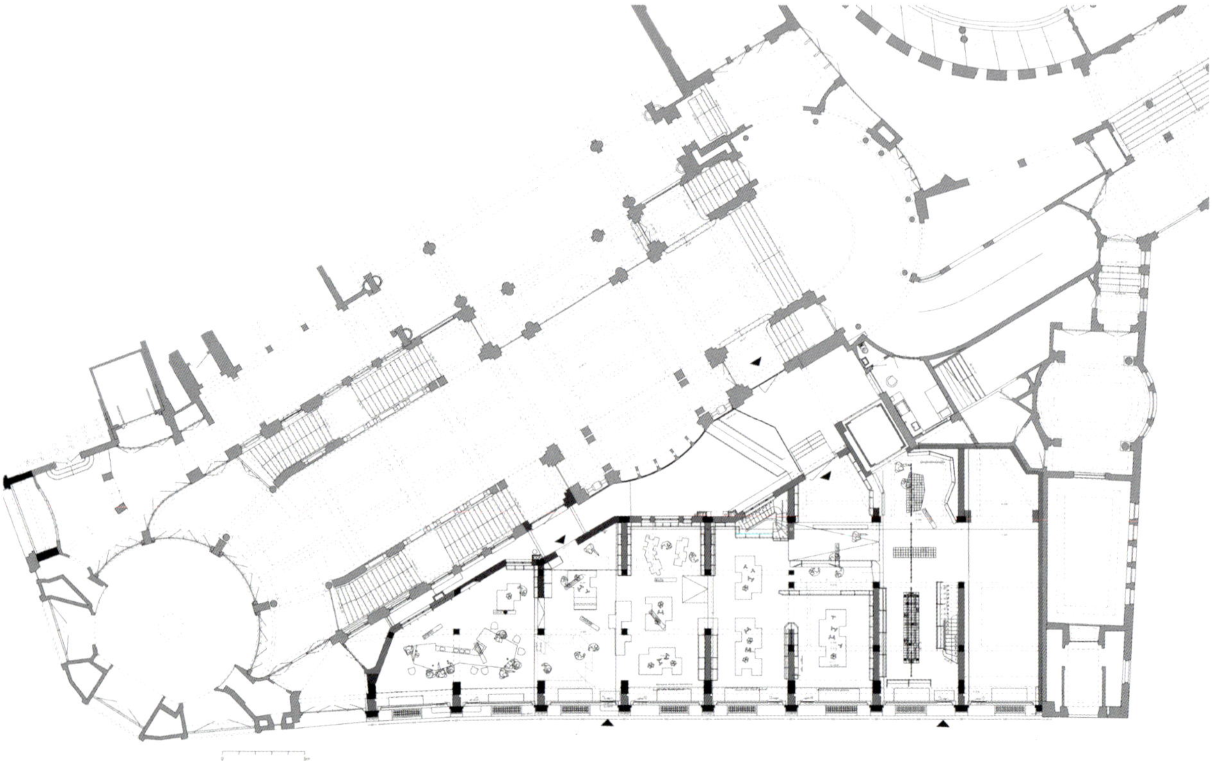

GENERALI
ALICE
CHRZ
LE CHATEAU

CUCCHI
Emilio Trad
LEE UFAN
ART

Calibre clothing little collins street
Melbourne, Austrlia

凯丽博莱时装店
澳大利亚，墨尔本

Calibre clothing is a Melbourne based menswear company and applys itself to hip men’s fashion. All projects are client specific and are given an individual response. The designer succeeds in creating a retail environment that reflects Calibre's quality and glamour.

Materials were selected for warmth and contrast and used in unique ways. Each slab was cut and laid in custom pattern. The recycled floor boards serve as wall lining and the mirror framing creates an elegant touch. The materials are either recycled or of low impact to the environment. The architectural form of the plaster ceiling is highlighted with a crisp white tone allowing for the sculptural three dimensional design of it to become more graphic. The crisp white colour reflects lighting within the space and gives a sense of lightness and calm.

凯丽博莱是墨尔本的一家本土男装公司，致力于当今时尚男装。所有的项目都以特定客户为指向并给出了不同的各具特色的回应。设计师成功的营造了一个零售环境，以此来彰显凯丽博莱的品质和魅力。

柔和和反差是选择材料的标准，并且材料的使用都是以一种新颖独特的方式。每个板层都被切断，摆放成传统的图案，可循环利用的木制地板被用于炉壁内衬。所用材料是可回收的或者是对环境影响不大的。石膏天花板的建筑形式是采用一块白色的清脆的石头塑造的突出的效果，考虑到它的空间设计的三个层次变得更为形象。

PROJECT
Calibre clothing little collins street
DESIGNER
David Hicks
COMPLETION
2005
PHOTOGRAPHER
Earl Carter

项目
凯丽博莱时装店
设计师
David Hicks
完工时间
2005
摄影师
Earl Carter

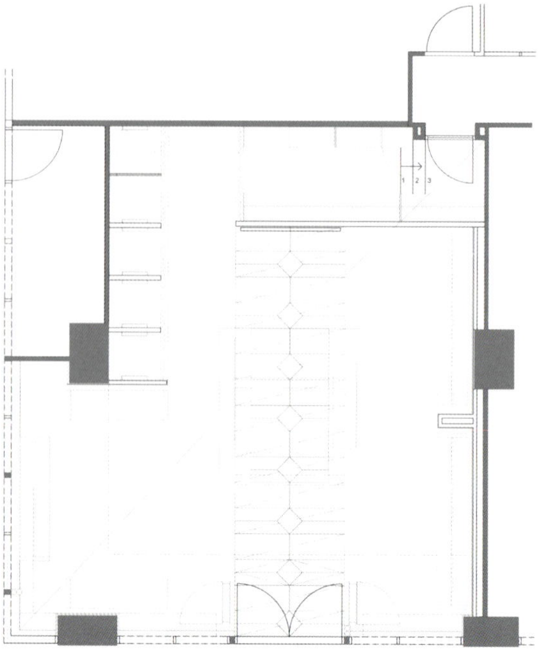

CIBOL in Beijing
Beijing, China

北京喜彪
中国，北京

Shop and Showroom of Ceramic Sanitary equipment manufacture company. Contrast of black and white, sanitary equipment is displayed as object of art in black space.

In comparison with other surrounding exhibition halls of the same function, the hall is inlaid with black boxes instead of white boxes as commonly used in galleries.

In a space where black prevails, the spot light illuminates the white sanitary wares suspending in the air. Such as it is, all the decorations are dimly visible and therefore weakened, revealing the one-on-one relationship between wares and man. In an unusually decorated space are displayed usual sanitary wares in the way of artworks. The relationship between wares and man turns the originally simple shopping behavior into wares appreciation.

陶瓷卫生用品的商店及展示厅，黑与白的对比,洁具仿若艺术品般展现于黑色空间中。

其他品牌的洁具店铺都被设计成“明快温馨的家庭型”风格，而我们的设计不是像美术馆追求的“白盒子”效果，而是做了一个“黑盒子”。

黑色的空间中沐浴、浮动在射灯下的白色洁具。去掉一切装饰后产生了“物体与观者，一对一的效果”。在另类的空间中，日常生活中最常见的洁具看上去就像艺术品一样。物体与人的关系也由购买变成鉴赏。

PROJECT
CIBOL in Beijing
DESIGNER
Keiichiro Sako,
Aya Unagami,
SAKO Architects
COMPLETION
2006
PHOTOGRAPHER
Seiichi Aoki (Beijing NDC Studio, Inc.)

项目
北京喜彪
设计师
Keiichiro Sako,
Aya Unagami,
SAKO Architects
完工时间
2006
摄影师
Seiichi Aoki (Beijing NDC Studio, Inc.)

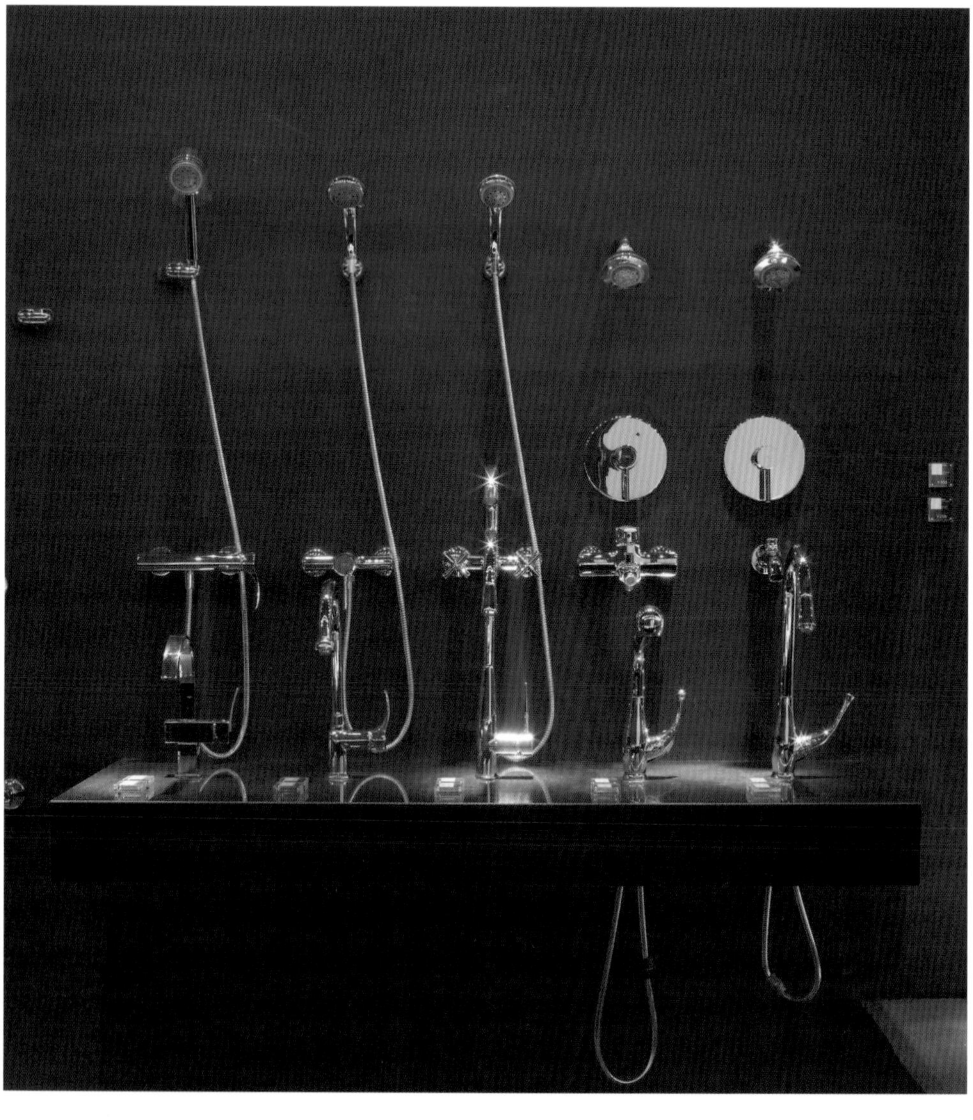

DC Shoes
New York, USA

DC 鞋店
美国，纽约

Teaming up with lifestyle giant, Quicksilver, DC Shoes was interested in developing their own retail identity. A famous design company, Graft was then asked to aid in this process by designing their Flagship store in the fashionable Soho district of NewYork City.

We can see architecture of new combinations, the crossing of different cultures. Unexpected and surprising misunderstandings, global transfers of architectonic beauty, robust crossbreeds - architecture derived out of circumstances, which can only be created through the grafting of different realities.

In turn, Graft provided the client with a spatial retail concept that would be a built expression of the company's values. This space would thus manifest into an innovative and unique interface for the DC Shoes customer to truly experience this counter cultures ideals.

DC Shoes与Quicksilver联手合作，并致力于发展他们自己的零售业。于是，Graft受DC鞋店委托，在纽约城最繁华地段设计了他们的零售旗舰店。

我们能够看到建筑的新式组合，不同文化的交织。出乎意料的不按常理的惊喜，国际流传的建筑美感，强大的异类同化——建筑源于环境，只有将不同的关系相连接才能创造出这样的效果。

Graft提供了零售的空间概念，这一设计体现出了DC鞋店公司的价值观念。从而在这个空间里，创造出了一个新颖独特的界面，让这里的顾客更真切地感受到这种反文化的理想。

PROJECT
DC Shoes
DESIGNER
Graft
COMPLETION
2006
PHOTOGRAPHER
Graft

项目
DC 鞋店

Graft
完工时间
2006
摄影师
Graft

DC
REMIX

CAVIAR
CARTEL

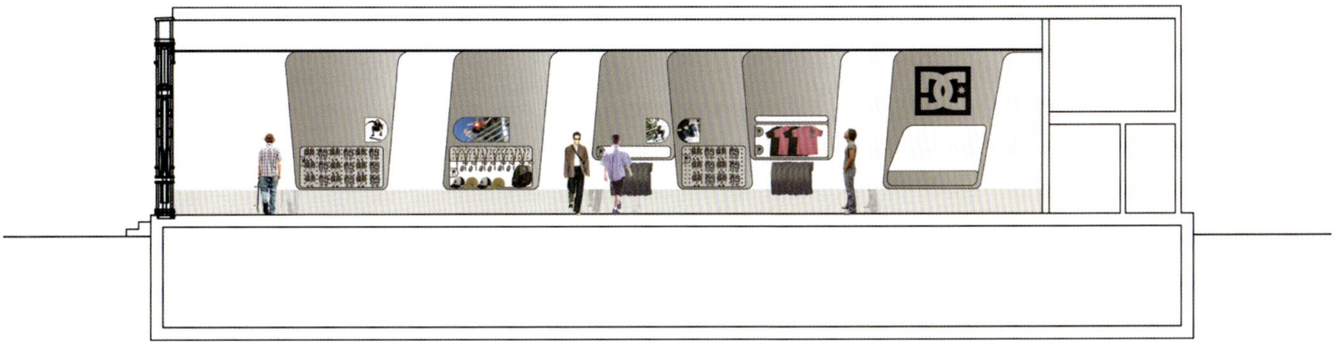

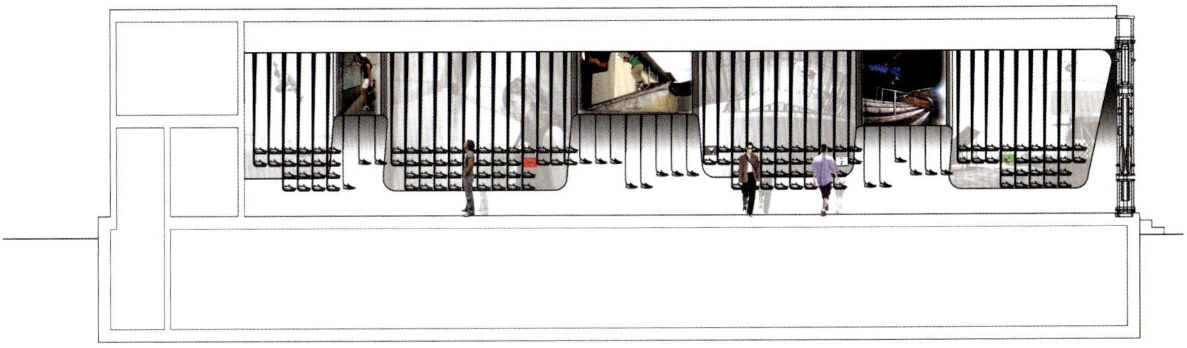

Decora Tokyo
Tokyo, Japan

东京装饰眼镜店
日本，东京

Decora Tokyo is an optical shop, opened in spring, 2007 at the Marui building.

The glasses' frames in this shop are almost in a same style, each of which is delicate and exquisite, therefore, the composition and the texture of the space build up the main visual factor. What's more, for the human being's restricted visualisation, if 2000 pairs of glasses were shown at the same time, it would cause the visual pressure contrarily. While by the changes of the depth which generated from the convex props, more space effects and migration would be guided.

Designer always grasp the trend making the new epoch in interior design scene. The spaces always give out the peculiar atmosphere not only in the space of its own but also in the periphery around it. The shops he designed always become to be recognised as the landmark and the cultural symbol of its town, and always the new prosper street would appear after it. It is because he is gifted with intuitively talent to find out the potential energy of the street.

Decora Tokyo是2007年春天在新丸井大楼中开业的眼镜店。

店内眼镜的镜架款式几乎全部相同，无论哪个都十分纤巧精致，因此，空间的构成和质感就组成了主要的视觉因子。另外，因为人类的可视范围被限制在一定区域内，如果将2000个眼镜一同展示出来的话，相反会造成视觉上的压迫感。设计师选择不直接展示商品，而是使用高低不同的道具，营造出更多的空间效果。

设计师往往抓住潮流趋势，并创造室内设计的新纪元。空间常常能够展现出一种特别的氛围，这种感觉不仅仅是空间本身所体现的，也体现在空间的边缘部分。设计师设计的店铺通常成为当地公认的地标性建筑，同时其中也蕴含了当地的文化气息，紧随其后还会诞生一条兴旺的商业街，这是由于设计师能够找到这条街区潜在的能量。

PROJECT
Decora Tokyo
DESIGNER
YOSHIHIKO MAMIYA
COMPLETION
2005
PHOTOGRAPHER
YOSHIHIKO MAMIYA

项目
东京装饰眼镜店
设计师
YOSHIHIKO MAMIYA
完工时间
2005
摄影师
YOSHIHIKO MAMIYA

décora
décora

décora

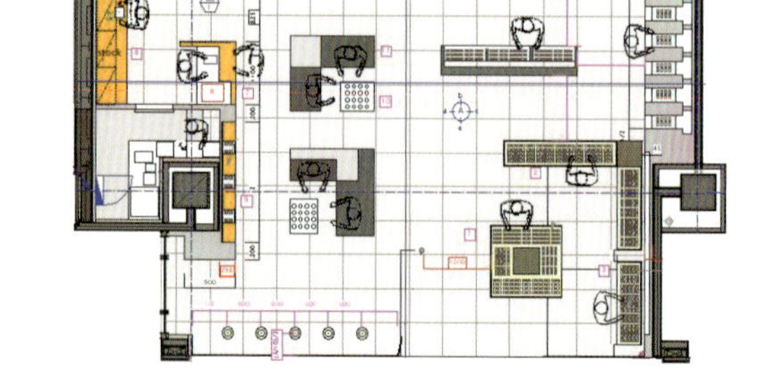

EXIT出口

DJS
HongKong, China

缀饰
中国，香港

DJS is a new jewellery label launched by Chinese Arts & Crafts (HK) Ltd. selling diamond & jade. The interior & C.I. design strategy adopted was to make use of the common natures/ chemical structures & cut shape "facets" of diamond & jade to generate a unique identity for the brand.To achieve this, the interior design made use of the metaphor of interpreting the retail space as a glowing gemstone.

White spatial envelope of glowing alabaster showcase wall units and white marble floor were firstly set up to provide a minimal setting to the space. Symmetrical layout was then adopted to echo the squareness of the site. This was followed by combining the geometrical composition of facets in both aesthetic and functional layers.

“缀饰”为中艺（香港）有限公司旗下新的钻石、翡翠品牌。在室内和企业形象设计上，运用了钻石和翡翠不同形状的“琢面”和共同的性质、化学构造，突出品牌的独特形象。为了达到这种效果，在室内设计上运用比喻手法，把购物空间演绎成一颗“宝玉”。

环境为绕亮白色的条纹大理石壁橱，搭配白色大理石地板，为空间营造微细的基调；对称的布局与店铺的方正相呼应，再融合宝石琢面的几何构造，美观和实用兼备。

PROJECT
DJS
DESIGNER
PANORAMA International Ltd.
COMPLETION
2008
PHOTOGRAPHER
Ng Siu Fung

项目
缀饰
设计师
泛纳国际设计顾问有限公司
完工时间
2008
摄影师
Ng Siu Fung

White glowing wall units with tilted vertical faces in random angles allow green side facings to be exposed. Small and precious items of diamond and jade displayed on custom-made frosted acrylic display stands were front and back-lit within each showcase and sculptural island unit to give sparkling and translucent beauty respectively.

The interplay of angular language is finally completed with the slanted mirror stainless steel ceiling to give distorted mirror images and further enhanced sculptural quality of the space which projected strong visual impact through the full-height clear glass shop front.

亮白石壁橱有不规则的纵面，露出翠绿面。这种微妙的绿色点缀，道出了翡翠色泽玲珑剔透的自然之美。方块内，特制磨砂亚克力展示架上，灯光前后辉映，小巧精致的钻石、翡翠、岛状雕塑晶莹剔透。

店铺空间以棱角为整体风格，倾斜的镜面不锈钢天花板映照出扭曲的镜像，增添店铺空间的雕刻美，再透过店面落地玻璃，投射出强烈的视觉冲击。

EXIT

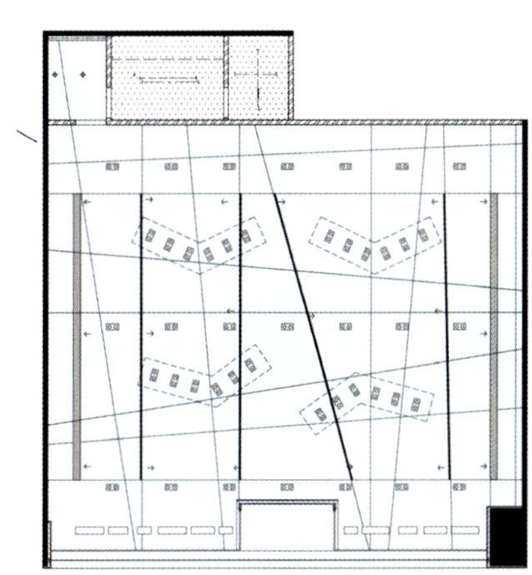

FELISSIMO in Beijing

Beijing, China

北京芬理希梦

中国，北京

We developed a very flexible unit furniture termed "cubic furniture" by considering boxes with two sizes (400 mm x 400 mm x 400 mm and 400 mm x 400 mm x 800 mm) as the basic sizes. By creating one or two round holes on the four sides of the box as the key design point, a single box can be connected from the top and bottom as well as the left and right sides through these round holes. They can also be assembled like a beam, on which both a mirror and a signage can be fixed and a power cord can be connected. It can also be made as a clothes rack if a steel tube is installed through the holes. Various functions can be realised through these simple round holes.

When evening sets, what attracts attention is the numerous neon droplights observed through the glass windows. The shop appears like a "forest" of white boxes surrounded by white light. Simple modeling creates a diversified world. Shoppers can choose goods that appeal to them while wandering through the white forest.

以40cm x 40cm x 40cm、40cm x 40cm x 80cm两种规格的箱子为基本元素的组合式家具"Cubic Furniture"的方案。箱子的四个侧面上开有1个或2个圆孔。这些圆孔是此次设计的重点。单个的箱子通过这些圆孔可以上下左右相连，也可以像梁那样悬挑出去。还可以固定镜子、标识，也可以贯穿电源线。箱子之间架上钢管就可以做衣架。通过简单的圆孔实现了各种各样的功能。

到了晚上，映入眼帘的是反射到玻璃窗上的数不清的霓虹吊灯，整个店铺宛如一个被白色光芒包裹的白色箱子的森林。简洁的造型创造出来的多样世界。客人徜徉在白色的"森林"中的同时，期待着找到心仪的商品。

PROJECT
FELISSIMO in Beijing
DESIGNER
Keiichiro Sako,
SAKO Architects
COMPLETION
2005
PHOTOGRAPHER
Minoru Iwasaki

项目
北京芬理希梦
设计师
Keiichiro Sako,
SAKO Architects
完工时间
2005
摄影师
Minoru Iwasaki

FELISSIMO

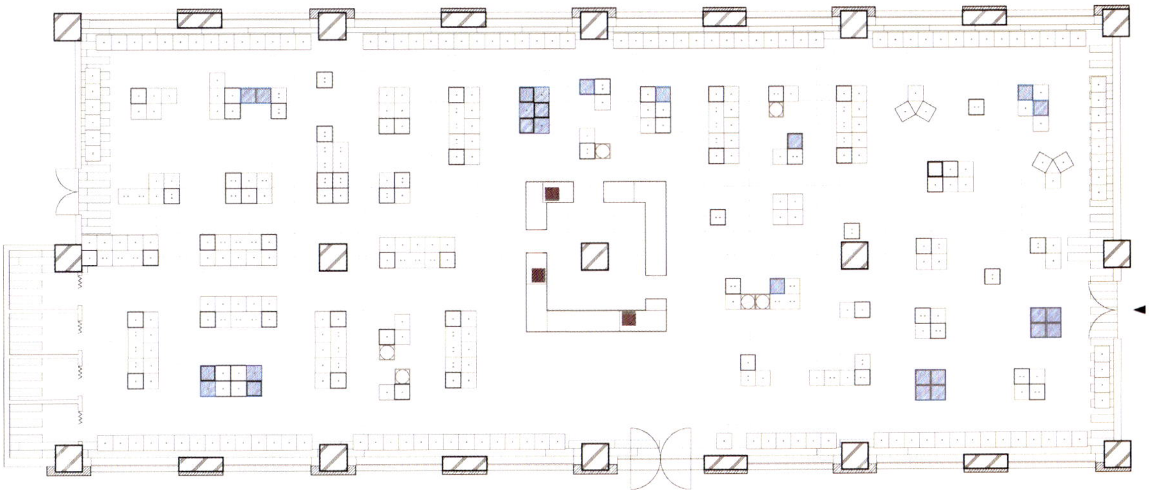

FELISSIMO 2 in Beijing

Beijing, China

北京芬理希梦 2

中国，北京

At FELISSIMO people want to make sure that the needs of the individual at present, speak to the needs of the world in the future. Their design goal is to celebrate the unique genius of every individual and use it to inspire others.

All that is used to display fashion dress and clothes is reduced to pits and plug-ins.

Up to 1000 pits are arranged on walls and on the floor. All furniture and decorations like clothes-hanger, tables, cabinets, flower vases and posts are inlaid with steel tube plug-ins which could freely be applied to the pits. The pits are connected to power supply, which makes it possible for lighting fixtures to work. The plug-in of the circular light tube is used to rotate the face-in plug-ins by 90 or 180 degrees, the manner of node is simple and clear. As a result, the space quality is greatly improved by simply altering the way finished products are laid out and displayed.

在芬理希梦，人们想要确定目前每个人个性化的需求，并与未来国际化的需求对话。他的设计目标是，弘扬独一无二的个性化天赋并以之鼓舞他人。

把时装店所必需的器具还原成“圆孔”和“管子”。

在地面、墙面上分块打了1000个圆孔。用铁棒做成的衣架、桌子、隔板、花瓶、标识等，可以随意插在1000个圆孔中。圆孔中预留了电线，照明也可以随意布置。圆形荧光灯管的连接部分做了简单的细节设计，即将通常朝着内侧的接线部分旋转90度或180度。即使是最普通的东西，稍稍改变它的摆放、安装方法，不是也可以改变空间的品质吗?

PROJECT
FELISSIMO 2 in Beijing
DESIGNER
Keiichiro Sako,
Shuhei Aoyama,
SAKO Architects
COMPLETION
2006
PHOTOGRAPHER
Seiichi Aoki (Beijing NDC Studio, Inc.)

项目
北京芬理希梦 2
设计师
Keiichiro Sako,
Shuhei Aoyama,
SAKO Architects
完工时间
2006
摄影师
Seiichi Aoki (Beijing NDC Studio, Inc.)

FELISSIMO

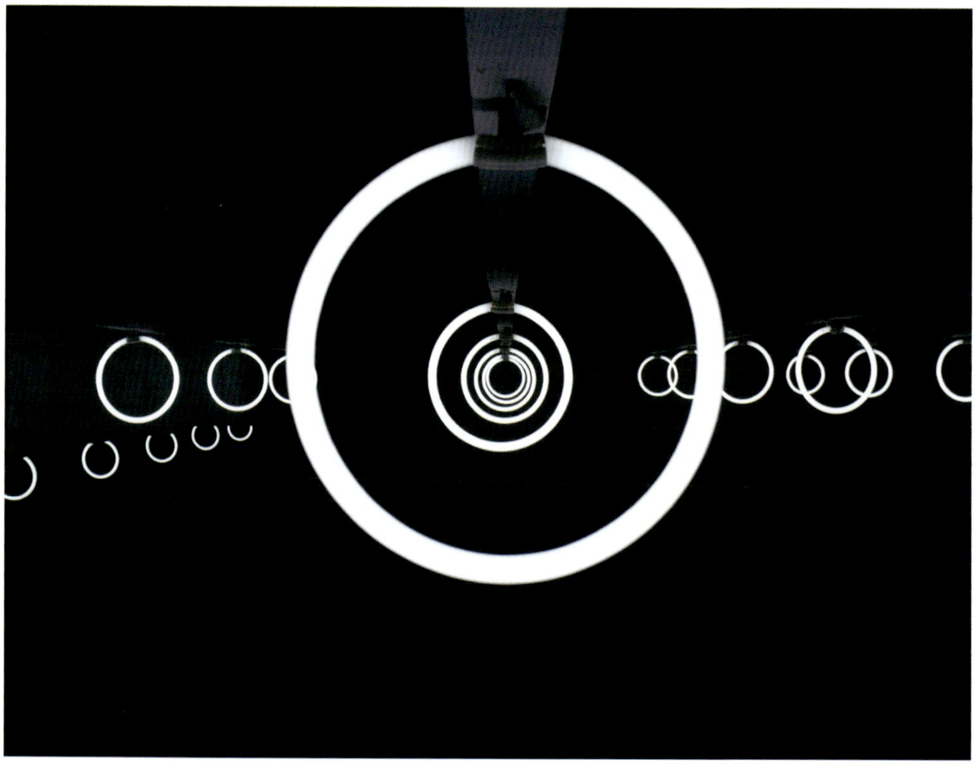

Fendi Flagship
Rome, Italy

芬迪旗舰店
意大利，罗马

The Fendi flagship store in Rome represents the re-launch of a luxury brand. Peter Marino was asked to design a store concept which emphasizes the brand's Roman identity both in a local sense as well as in an extended and universal sense of Rome as caput mundi. The store has multiple entrances allowing pedestrians access to the various passages and rooms of the store which occupies two floors of the Palazzo Boncompagni Ludovisi (an early 20th century neo-classical residence with ground floor shops).

Rome was identified in the material essences used, such as travertine and the volcanic cobble stones called sampietrini. Travertine was represented in tactile and visual terms as horizontal striations. And, the encrusted rust often found on the surface of Rome's travertine fountains also present another Roman essence. Variations of this theme can be seen in irregular ribbing of surfaces encrusted in rust, in basalt, and amber glass.

芬迪在罗马的旗舰店意味着这个豪华品牌的重新发布。店主要求彼得·马里诺设计一个概念店，从两个方面强调该品牌在罗马的地位，包括对本地的意义，以及罗马发展壮大成“世界旗舰”的意义。商店在路德维希宫殿（20世纪初的新古典主义住宅，一楼是商铺）占有两层楼的位置，设有多个出入口，让过路的行人能直接进入店内的各条通道和各色房间。

罗马擅长充分利用材料的内涵，如钙华和火山卵石石块等，这种石块被称为“圣彼得广场上的小石头”。钙华看上去和摸起来都有横向条纹。而罗马钙华喷泉经常出现的斑斑锈迹也代表了另一种罗马的精髓。店内的设计对这一特色加以改进，在不规则的墙面上，除了锈迹还会饰有玄武岩和琥珀玻璃。

PROJECT
Fendi Flagship
DESIGNER
Peter Marino Architect
COMPLETION
2005
PHOTOGRAPHER
Matteo Piazza

项目
芬迪旗舰店
设计师
彼得·马里诺建筑公司
完工时间
2005
摄影师
马特奥·皮尔萨

Yellow travertine is used as a flat wall surface, as well as a curvilinear valence floating above the merchandise racks. In the fur salons rust encrusted striations serve as walls below the travertine curves, while handbags float on thassos shelves and in niches cut into the yellow travertine. Here white striated curves float above the goods. Shoes rest on travertine planes floating below curved travertine curtains.

The dark gray metallic restoration of the neoclassical stair becomes a counterpoint to the chromatic richness of the store interiors. Lighting is specific to each space in the interior and exterior.

Fendi's identity is expressed through non-standard architecture in which walls become indeterminate shells moving through the spaces they create. The brand is no longer a fixed; instead here it becomes dynamic and inventive.

平直的墙面和商品陈列架上方的曲线板上都使用了黄色的钙华。在皮草区，墙面上是锈色条纹，上方则是钙华曲线板，理石台板直接嵌入黄色钙华墙面里，台板上放着的手袋犹如悬在半空中。白色条纹的曲线板浮在商品上方。鞋子静静地立在钙华展台上，上方也有弯曲的钙华帷幕。

复古的深灰色金属楼梯与色彩丰富的店面设计形成对比。店内外各个区域的照明设计也各具特色。

芬迪的品牌形象，通过不规则的设计结构展现出来，墙壁成了难以捉摸的轮廓，在设计师创造的空间里流动。品牌不再一成不变，而是充满了活力与创造性。

CASH
WRAP

HUGO BOSS
Shanghai, China

HUGO BOSS男装店
中国，上海

BOSS Orange offers casual collections for men and women. The interior of the store is a contrast-rich interplay of authentic natural materials combined with high-gloss paint, white finished metal and brass surfaces. This combination creates an impact contrast for "radical chic". Coloured walls with finely etched graphics form a beautifully accented impression.

The ground floor functions with his two entranceways and the 65sqm large surface area, as an opener to the ground floor. After walking up the stairs, with his wire netting handrail, you will get to the first floor, were the two fashion boxes with the integrated changing rooms and a lounge area with vintage furniture. Hanging racks enframe the room. Affectionate details which you will notice on the furniture and on the walls, the raggedness and the individuality transports the image of the street and the fashion inside.

BOSS Orange店专为时尚男女提供休闲收藏品。内部设计将纯天然物质与高光喷绘、白色成品金属及黄铜表面相结合，对比鲜明，视觉冲击感强烈，彰显别致。深色的墙壁与精雕细刻的图案，形成一幅绚烂美丽的风景画。

底层设有一个65平方米的大厅及两个走廊。顾客沿着铁丝网扶手拾阶而上，来到一楼。里面设有综合更衣室及配备老式家具的休息区，两个时尚包间。房间里挂满了晾衣架。无论在家具还是墙壁上，无处不呈现着商店内外的粗糙、天然与时尚个性的冲击。

PROJECT
HUGO BOSS
DESIGNER
RaiserLopesDesigners
COMPLETION
2006
PHOTOGRAPHER
Martin Grothmaak

项目
HUGO BOSS男装店
设计师
RaiserLopesDesigners
完工时间
2006
摄影师
Martin Grothmaak

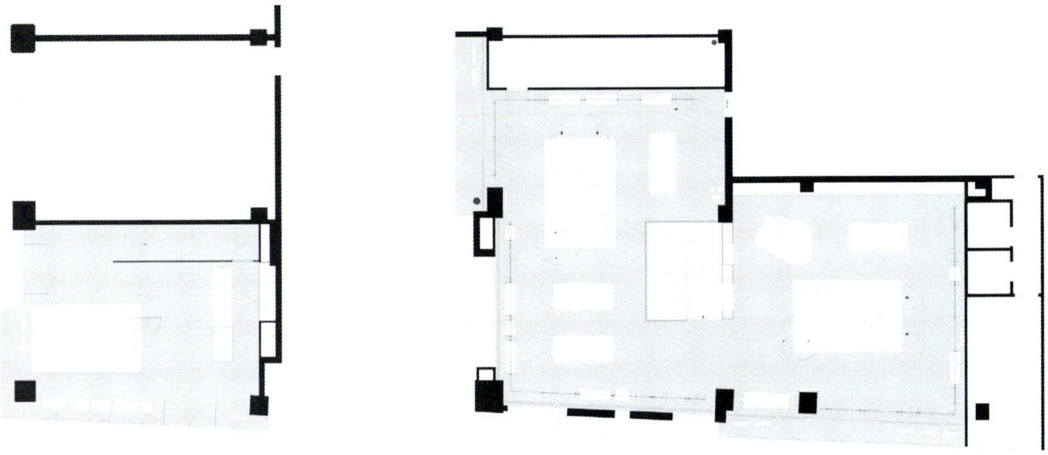

Hugo Boss Orange Store
Madrid, Spain

雨果·波士橙色店
西班牙，马德里

BOSS Orange offers casual collections for men and women. The interior of the store is a contrast-rich interplay of authentic natural materials combined with high-gloss paint, white finished metal and brass surfaces. This combination creates an impact contrast for "radical chic". Coloured walls with finely etched graphics form a beautifully accented impression.

The new store in Madrid though comes up with more than fashionable surprises. The room are structured by „fashion boxes", cubes made from maritime pine panels, in which the clothing s displayed.
Fashion boxes:
Fashion boxes which are placed in the shop, provide a modulated room, different shop zones, like a lounge area and functional areas can be put together and be interlaced. Because of the individual construction of the boxes in form and function every supplied space can be upgraded.
The tension between the inside and outside which is drawn throughout the concept, enables a wide range of unexpected surprising elements for the customers.

波士橙色店里同时出售男装和女装。店内，强烈的反差处处可见，天然材料配以高光泽涂料，白色的成品金属表面却涂着黄铜色。这样的组合给“激进的时尚”带来了冲击。彩色墙面上精雕细琢的图案形成了美丽的异乡情调。

马德里的新店面带给人们的惊喜，已远远超过了时装本身。展示区由“时装方块”构成，成衣都放在海松板做成的大方块上出售。
时装方块：
时装方块让商店变得与众不同，像是休闲区，其实又是功能区，二者交融在一起。
每个方块的大小功能各不相同，起到了提升空间的作用。
店内与店外的紧张关系贯穿整个设计理念，为顾客带来各种各样意想不到的惊喜。

PROJECT
Hugo Boss Orange Store
DESIGNER
RaiserLopesDesigners
COMPLETION
2007
PHOTOGRAPHER
Hartmut A. Raiser

项目
雨果·波士橙色店
设计师
赖泽和洛佩兹设计师
完工时间
2007
摄影师
哈特穆特·赖泽

EXTINTOR DE
INCENDIOS
REBAJAS

BOSS
REBAJAS
30%

REBAJAS

BOSS

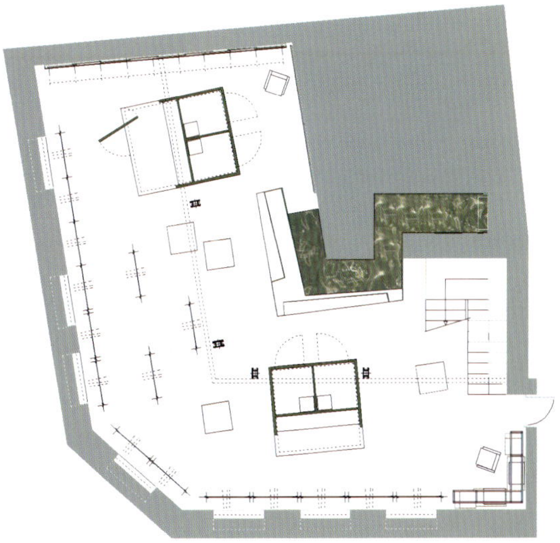

Italian gelato RONO

Nagoya, Japan

RONO意大利雪糕店

日本，名古屋

It let you imagine Italy with the gelato.

Designers create this unusual interior for an ice-cream shop. The whole interior is deliberately characterless, with no colour, texture or reflections, so there is nothing between the customers and the happy colour in the ice-cream tubs.

It is plainness as the icon of the shop and originality to have keept in mind in this design. There are two reasons about using a muted color palette. One is to enhance the color that the material itself of the gelato. Another reason is because I wanted to express the unrealistic element.
The silver cross on the counter-top is a metaphor as an unrealistic thing for the Japanese at the same time as the cross which was put without difficulty is a device to express the Western world like the graphic of the glass and the mirror.

这个设计让你联想到意大利冰淇淋。

这个雪糕店是设计者不同寻常的室内设计。整个室内设计刻意展现出毫无特点的风格，没有缤纷的色彩、质感和映像。所以在顾客和色彩灵动的冰淇淋筒之间没有任何阻隔。

直白的商店标志创新让人记住这个设计。使用暗色板有两个原因，第一是想突出材料本身的色彩，另一个是设计师想诠释一个不切实际的结构。这个在柜台顶端的银色交叉线来暗喻日本人不符合实际的东西，它总是很容易的放在那里。这个设计表达西方人喜欢平面玻璃和镜子。它的用途是来挡住包装好的冰淇淋盒子。

PROJECT
Ialian gelato RONO
DESIGNER
HIROYUKI MIYAKE
COMPLETION
2008
PHOTOGRAPHER
RIKOH ADACHI

项目
RONO意大利雪糕店
设计师
HIROYUKI MIYAKE
完工时间
2008
摄影师
RIKOH ADACHI

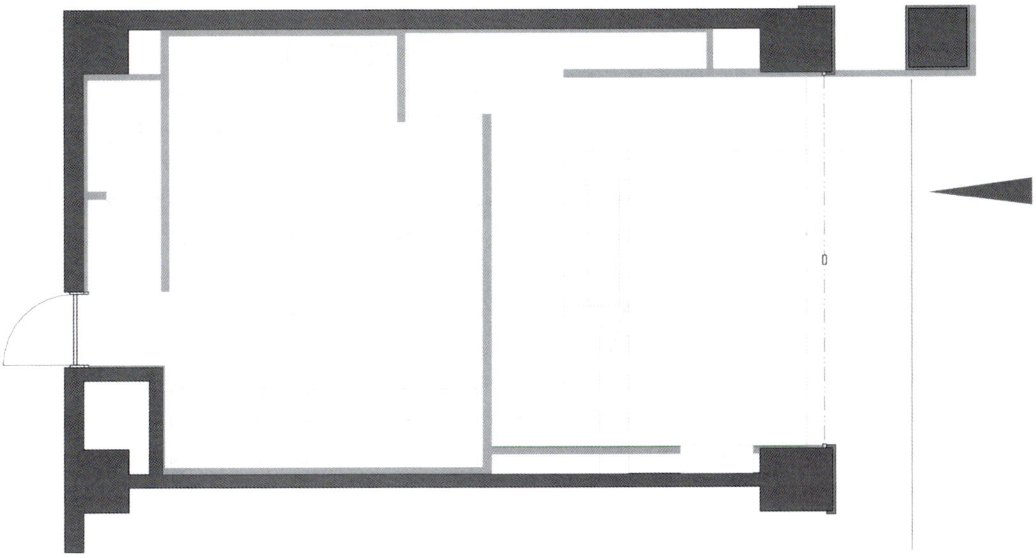

RONÓ

RONÓ
Italian gelato
RONÓ

RONÓ

++ DER AUGAPFEL
ER GEFUELLT DIE DIE
EL SEINE FORM BEHAELT
ES MAEDCHEN PUEPPCHEN
H DIE DAS LICHT EINDRINGT
CHTSTRAHLEN DIE INS AUGE
FLEXION ++ INS AUGE GELANGEN
WIR SEHEN

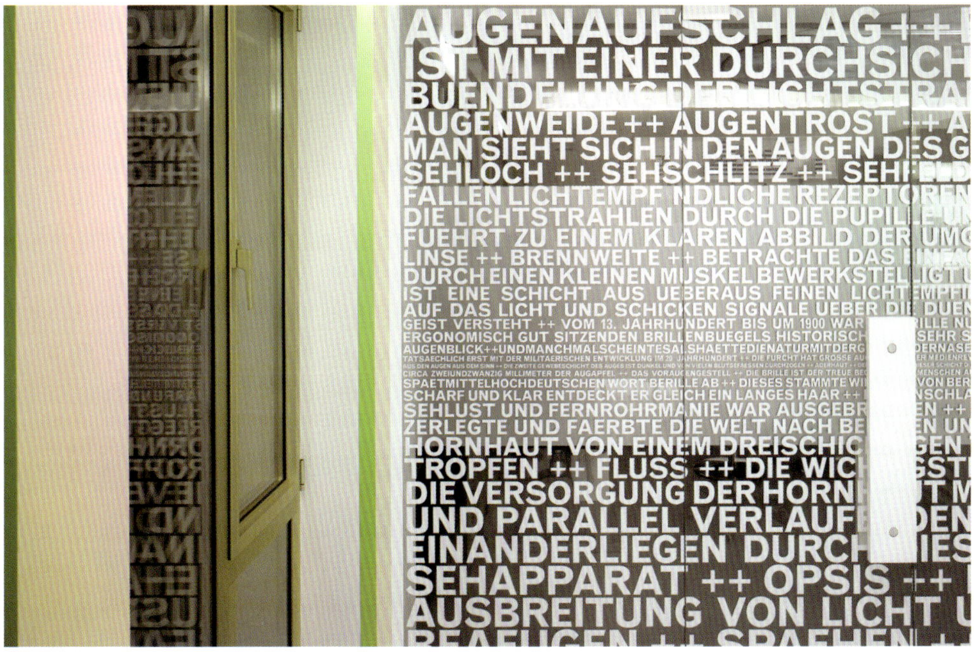

Kaestner Optik Shop

Stuttgart,Germany

凯斯纳眼镜店

德国，斯图加特

Kaestner Optik is an optometrist's shop with a long tradition of excellence, situated in the very heart of Stuttgart.

Two distinctive key visuals are elaborated in the new Optical Center: first, the stunning view through the colonnades onto the Schlossplatz, and second, an oversize, blue iris on the ceiling with a diameter of four meters.

The space is taken in two visual directions. The curved upper corners of the room, the convex and molded iris create a kind of tunnel effect, which shifts the view out of the window to the central stage. A comb-shaped, back-lighted shelf running from the stairway and the upholstered long wall that lies behind it also support this shift. At the same time, the room is oriented around the long glass wall on the opposite side of the room, which screens the more technical refraction rooms and contact lens-fitting area that lies behind it.

凯斯纳是一家历史悠久的优质眼镜店，坐落在斯图加特市中心。

在这家新的光学中心有两个值得说明的、与众不同的视觉重点：第一，通过石廊柱到中央大厅有极漂亮的视角；第二，还有一个巨大的直径4米由蓝色虹膜包裹的圆形天花板。

这一空间有两个视觉方向，房间上部的弯曲的拐角，房间内凸起的虹膜形成了一种隧道效应，将窗外的景色转到室内来。背部照明的梳形货架从楼梯及楼梯后的装饰长墙延伸过来。同时，对面的长玻璃墙放大了房间的空间感，玻璃映照出后面的技术屈光室和验光室。

PROJECT
Kaestner Optik Shop
DESIGNER
ippolito fleitz group
COMPLETION
2007
PHOTOGRAPHER
Zooey Braun

项目
凯斯纳眼镜店
设计师
ippolito fleitz group
完工时间
2007
摄影师
Zooey Braun

DU SIEHST ++
+ DER AUGAPFEL
DAS STROM EIN BERE
OPTIK IST
STAMMEND VOM GRIECH
++ WIR SEH
++ DU SIEHST

AUGAPFEL
JELLT DIE DIE
FORM BEHAELT
HEN PUEPPCHEN
LICHT EINDRINGT
ILEN DIE INS AUGE
NS AUGE GELANGEN
ICHTSTRAHLEN UND
CHUNG ++ KONTAKT
AKKOMMODATION WIRD
WIRFT ++ DIE NETZHAUT
TOREZEPTOREN REAGIEREN
ESELLSCHAFT ++ EINE REGELRECHTE
LINSE IN DAS AUGE ++ MAN
CHE ZUR LUFT HIN IST DIE
S FEUCHT ++ DRUESE
AUTPARTIKEL UND
ST AUS DUENNEN
UZWEISE UEBER
E ++ SEHHILFE
SICH MIT DER
T ++ BLICKEN
- LEHRE VOM
UFSCHLAG

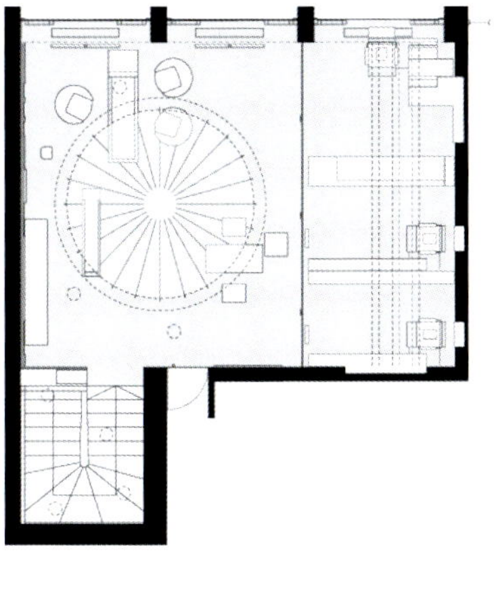

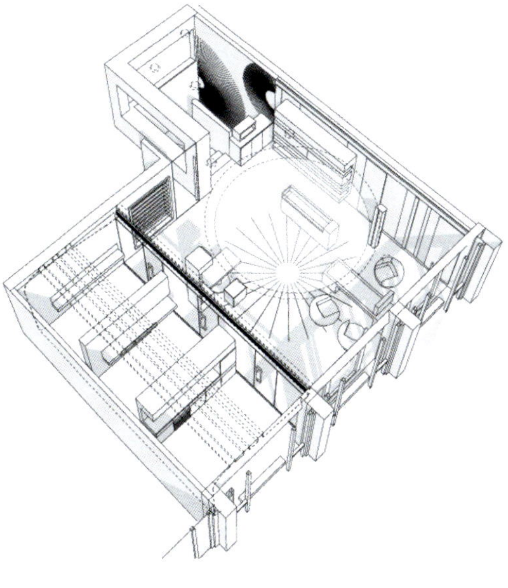

Kaloo
ATHENS

Kaloo
HongKong,China

卡路
中国，香港

Kaloo is an international brand, which was created in 1998 by a French national Mr. Eric Panciulo. This brand was distributed into 35 cities worldwide, such as Paris, Munich, Madrid, Istanbul, Tokyo, Hong Kong, Taiwan etc.

In respecting to the original spirit of the Kaloo, and adopting the European culture, the concept was inspired by " Love, Care, and Happiness "

Because the design would be implemented worldwide, our given objective was to create a retail image with flexible system, which to be easy adopted and easy appreciated by customer. Therefore, a warm, homely, harmony and welcoming atmosphere have been reflected by a modern way of mixing use of French Oak, and white wall. The design also integrated with 2 parties, modular system for easy and flexible showing Kaloo product, such as soft toys, furniture, and textile items. And soft elements have been created for images enhancement. In the Paris shop, we especially created a very homely look with creating Kaloo label furniture, which harmony integrated into a traditional French interior.

卡路是一个国际品牌，由法国人埃里克·潘西罗先生创立于1998年。该品牌遍布全球的35个城市，包括巴黎、慕尼黑、马德里、伊斯坦布尔、东京、香港、台湾等。

设计理念是“关爱与幸福”，尊重卡路的品牌精神的同时，融入了欧洲文化。

由于设计将在全球范围内实施，设计师的目标是建立一个灵活的零售店形象，容易被不同的客户接受。设计师采用法国橡木和白色的墙壁，营造了温馨、亲切、和谐、热情的氛围。店内还安排了两间展示室，方便和灵活地显示卡路的产品，包括毛绒玩具、家具、纺织品等。同时以布艺装饰强调品牌形象。在巴黎分店，设计师特意将店面设计成家的样子，以展示卡路的家具，并在店内和谐地融入了传统的法国风格。

PROJECT
Kaloo
DESIGNER
Axiom Oval Partnership Ltd.
COMPLETION
2007
PHOTOGRAPHER
Axiom Oval Partnership Ltd.

项目
卡路
设计师
Axiom Oval Partnership Ltd.
完工时间
2007
摄影师
Axiom Oval Partnership Ltd.

Kaloo

Kaloo

Kaloo
ATHENS

Kaloo

Lanvin Boutique Ginza
Tokyo, Japan

浪凡精品店
日本，东京

"Lanvin Boutique", located in Ginza Central Street (one of the busiest fashion district in downtown Tokyo), is a flagship shop of a French fashion brand that has a long history in couturier maison. We were chosen by the brand's designer, Alber Elbaz, as the architect for interior and exterior of the shop. Our design was inspired by a contemporary French maison (residence) where the modern and the classical meet. We proposed a façade of continuous steel plate with 3000 clear acrylic cylinders inserted as windows, to look like a Lanvin' s party dress on which diamonds are embroidered.
We typically use window frames, sealants, and many other materials for window details. However, we thought that any ordinary window details would ruin the subtle but yet very rich phenomenon that ephemeral light creates right at the openings. Borrowing from the craft of boat builders, we invented "hiyashi hame", a method to avoid having any obstructions at the window edge, aiming to turn the windows into something more than windows (we are now applying for the patent). We thought that the method would transform them into photons.

"浪凡精品店" 坐落在银座中央大街（在东京市中心一个最繁忙的时装区），是一个法国时装品牌旗舰店，在couturier maison有着悠久的历史。
我们选择的品牌的设计师Alber Elbaz来设计店铺内部和外部。我们的设计灵感来自当代法国之家（住宅），有着非凡现代和古典的接触。包在窗上的是一个打有3000个圆孔的长铁板的门脸，看起来很像浪凡派对装上面镶的钻石刺绣。
我们通常使用的窗框、密封胶以及其他许多细节材料来弄窗户。不过，我们认为，普通的玻璃窗质地薄而且易破损，我们认为，任何普通窗口微妙的细节将产生内容十分丰富的现象，开始创造短暂、轻微、直接的。从手工业造船者中借鉴，我们发明了"hiyashi hame"方法，以避免有任何障碍，目的是使窗户蕴涵更多的意义（我们现正申请专利）。我们认为该方法将转变成光子。

PROJECT
Lanvin Boutique Ginza
DESIGNER
NAP Architects
COMPLETION
2005
PHOTOGRAPHER
MAP Studio

项目
浪凡精品店
设计师
NAP Architects
完工时间
2005
摄影师
MAP Studio

LANVIN
PARIS

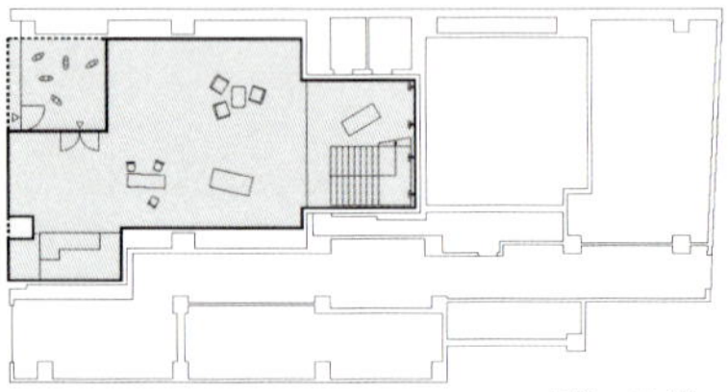

1F Plan 1 : 300

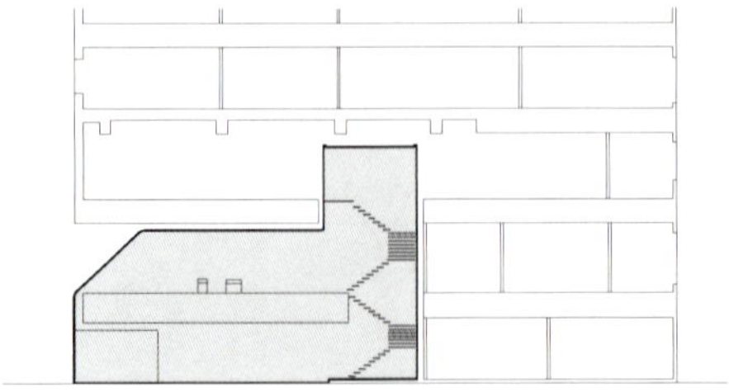

Section 1 : 300

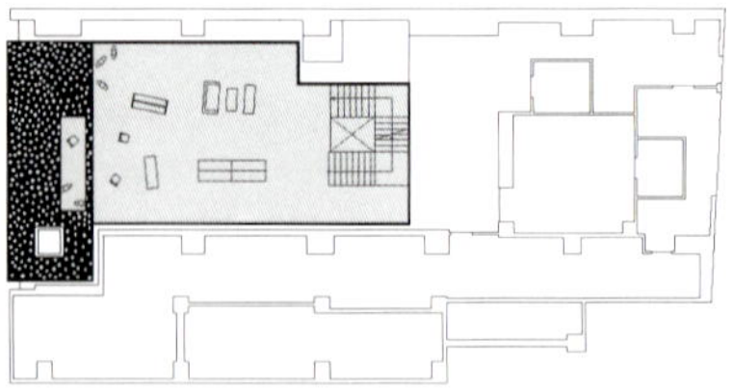

2F Plan 1 : 300

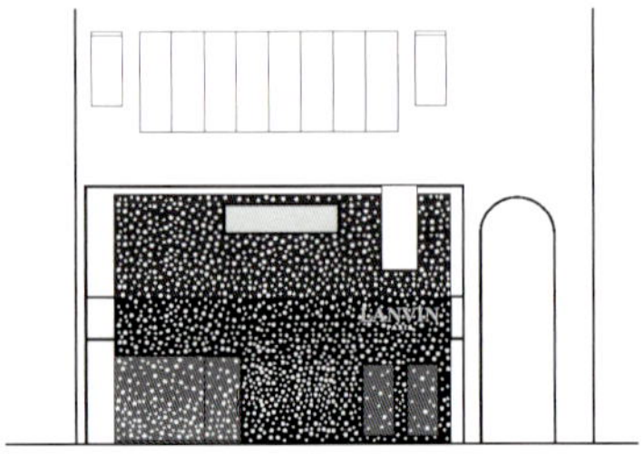

Elevation 1 : 200

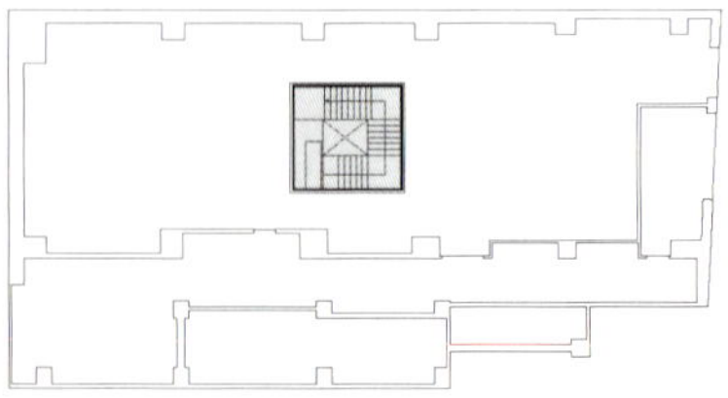

3F Plan 1 : 300

JL 704
QF 330
CX 305
EK 303

Louis Vuitton Flagship

Hong Kong, China

路易·威登旗舰店

中国，香港

The new three-level luxury boutique at the Landmark occupies a prominent corner in the central district of Hong Kong. The space is a twenty-five foot high semi-transparent box.

The first floor slab has been disconnected from the façade allowing a continuous 6.8M tall void along the street edge of the store. This allows added natural light to be filtered through the 'louvered' façade which is always present from the inside providing a shimmering background to the interior spaces.

The main stair volume, designed as if carved out of a solid block of stone, organizes the project by connecting three distinct levels. The sandblasted glass treads of the stair feature built-in L.E.D. panels containing video images on each walking surface. Customers can experience these video images individually with each step. Customers at ground level can also view the reflected images from the mirror on the ceiling, while the entire visual narrative spreading over three sections can be viewed from the uppermost level of the boutique. The fine jewelry store within the boutique can be accessed from a separate entrance and contains its own suspended stainless steel staircase made by stainless steel bars and stone treads sits.

这家新的三层豪华精品店位于香港中区置地广场的一个转角处，十分显眼。空间是一个25英尺高的半透明方块。

一层楼板被截断，在商店邻街的一边形成了挑空6.8米的无障碍空间。这让自然光透过“百叶”店身倾洒进来，给空间内部营造了闪烁的背景。

主楼梯被设计得如同从坚硬的石块上切割下来一般，贯穿三个不同的楼层。楼梯的喷砂玻璃踏板中装有LED面板，每块踏板中都有视频影像。顾客每踏上一个踏板都能看到单独的视频影像。一楼的顾客还能通过天花板上的镜面看到这些影像，一旦影像蔓延到三楼，就能俯瞰整个精品店。店内的珠宝店有单独的入口和不锈钢旋转楼梯。楼梯由不锈钢条和石质踏板组成。

PROJECT
Louis Vuitton Flagship
DESIGNER
Peter Marino Architect
COMPLETION
2006
PHOTOGRAPHER
Vincent Knapp, Stéphane Muratet, Peter Marino Architect

项目
路易·威登旗舰店
设计师
彼得·马里诺建筑公司
完工时间
2006
摄影师
文森特·科奈普，斯特凡·米拉泰，建筑师彼得·马里诺

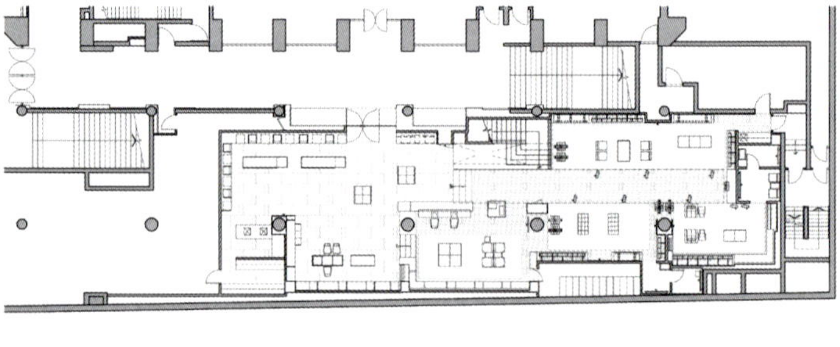

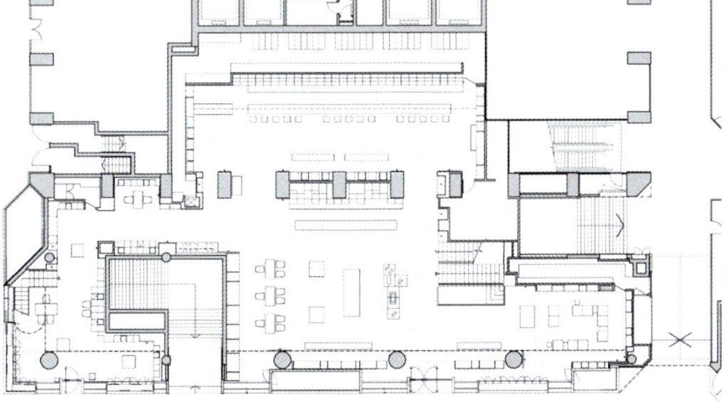

広告社
LUNCH VACATION
LUNCH
VACATION
注意

Lunch Vacation
Tokyo, Japan

午餐假日
日本，东京

Lunch Vacation is a new brand created by a famous Japanese company. Lunch Vacation is known for its fresh and delicious Bento "lunch boxes".

The most important thing for the designer is to design new concepts, new atmospheres which will give emotions to people who visit or use the space. For creating spaces using colours, or it is better to say the colours which used to create the space. The designer uses colours as three-dimensional items, like layers, in order to create spaces, not as an element applied after in order to complete or decorate the space.

We designed the interior and exterior of their very first shop opened in Kanda, Tokyo. As opposed to the typical Bento shop -rather dark and humble, we created a brand new image which targets "women" as their main customers; colours representing a variety of lunch box menu and the fine materials giving a healthy ingredients image. After the Kanda shop has opened, we designed the second and third shop which later opened in Tokyo.

午餐假日是日本一家知名公司创建的一个新品牌，因其美味可口的盒饭而著称。

对于设计者来说，最重要的是设计出新鲜的理念和氛围，以此来激发到访者和使用者的感情。为了装点空间而使用色彩，倒不如说是色彩营造了空间。设计者用色彩制造出三维立体的层次感，来打造空间效果，而不是将色彩作为一个完成并装点空间的基本元素来使用。

与那些典型的盒饭店不同，设计师打破了以往暗淡、简朴的环境，而是以女性顾客为主设计了一个新的品牌形象。不同的颜色带包不同的盒饭，优质的材料更增添了健康的气息。自这家店开张之后，第二家和第三家店也相继在东京落户。

PROJECT
Lunch Vacation
DESIGNER
Emmanuelle Moureaux Architecture & Design
COMPLETION
2005-2006
PHOTOGRAPHER
Hidehiko Nagaishi

项目
午餐假日
设计师
Emmanuelle Moureaux Architecture & Design
完工时间
2005-2006
摄影师
Hidehiko Nagaishi

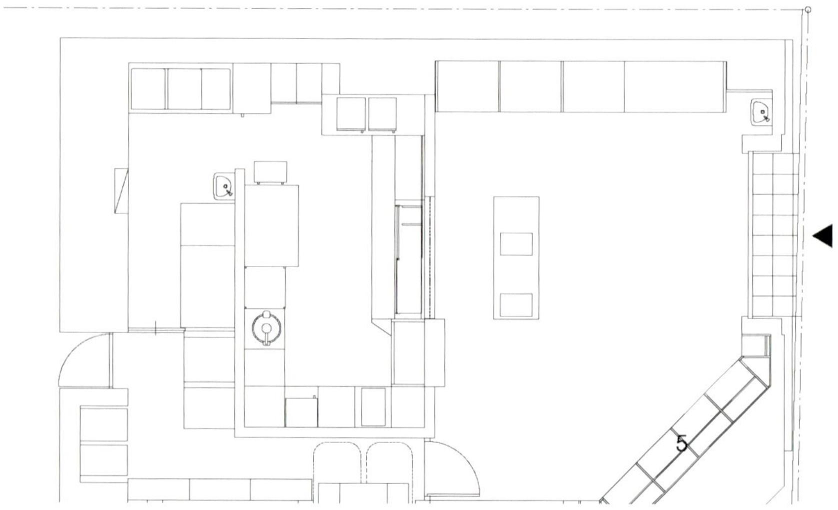
5

オーガニックホットコーヒー
¥120
エスプレッソタイプコーヒー
¥120
カフェ・ラテ
¥120
コーヒーはセルフ方式です。

Manner Shop
Vienna, Austria

风格巧克力店
奥地利，维也纳

The concept behind the design of the Manner shop on St. Stephen's Place unites the century-old history of a long-established business with the freshness of a modern, dynamic corporation.

The surface of little chocolate-brown tiles creates the impression of the whole area being coated in chocolate. This consistent dark colour causes the arches to lose their significance, thus allowing the products on display to stand out. Furthermore, the shiny surface of the small glass tiles creates an elegant, shimmering ambience with astonishing light reflections. Unusual,yet timeless.

In the style of the well-known Manner Schnitten, the shop interior also features a pink banderole on which the Manner merchandise can be presented. In this way the various product lines (which differ in colour as well as in form) may be associated with the brand name of Manner. Again analogous to the Manner Schnitten wrapping, the back of the presentation banderole has a metallic silver glaze which is backlit, thus creating a contrast to the room's dark mantle, and making it appear to be floating in space.Monitors placed among the products show pictures of the manufacturing process, as well as photos from Manner's past.

Manner shop 商店设计背后的概念是史蒂芬街，这个商店是汇聚百年历史的一项长期制定的生意和新鲜的充满生机的现代公司。店的表面的砖块有着巧克力一样的棕色，给人以涂上了一层巧克力的感觉。这种贯穿一致的深色使得拱门失去它的意义，使得店中的产品脱颖而出。此外，玻璃表面的小发光面给人以优雅的、闪烁的、奇异的反射光氛围。不寻常，但永恒。

众所周知，仿Manner Schnitten式，商店内部有Manner商品粉色飘带特征。不同的商品在不同的线上（不同颜色在不同的格子上）可以把各种品牌的东西与Manner品牌联系在一起。再加以类似于Manner Schnitten式的包装，包装背后展现出的金属银光是背光，因此与房间阴暗的顶棚形成对比，并使其似乎是漂浮在空中。商品中间的显示器展现商品的制作流程和商店过去的照片。

PROJECT
Manner Shop / Vienna
DESIGNER
Susanne Thomanek,
Thomas Duregger,
Robert Charuza
COMPLETION
2005
PHOTOGRAPHER
Rupert Steiner

项目
风格巧克力店
设计师
Susanne Thomanek,
Thomas Duregger,
Robert Charuza
完工时间
2005
摄影师
Rupert Steiner

Manner
ILDEFONSO
ILDEFONSO
ILDEFONSO
XL
XL

Manner
Manner
...mag man eben

Manner
Manner
Dekorhippen
Waffeltüte
Waffeltüte

SCHOKO-BANANEN
XL
Napoli
Dragee Keksi
neu

Drop Me a Line
USED JEANS

Me Boutique, Issey Miyake

Paris, France

三宅一生精品服饰

法国，巴黎

AEDS has worked with Issey Miyake for more than five years to create the Pleats Please spaces and me line boutique. The me line boutique is a practice in minimalism. Existing in an all white colour palette, the featured element of the boutique is a lacquered, reflective floor and ceiling with recessed light fixtures. This unique lighting method prevents direct spot light, creating an even, ambient light environment. The forthcoming fitting rooms, digitally designed and using 3-D print technology, will be an ornamental installation within the aesthetically pure me boutique space.

AEDS has matched this sentiment by using cutting edge digital technology and new materials to create the spaces that hold Miayke's designs. AEDS is always conscious of the fundamental necessity of creating a showcase environment for Miyake. The me line boutique exists as a platform to highlight and complement Miyake's designs.

设计师 Aeds 与设计师Issey Miyake合作了5年的时间去创作褶皱空间的Me 连锁服饰店。这个店惯用简约的风格。店中全部用白色墙板，特色是以光亮的漆面作板面的材料，所以可在地板和天花板反射出隐约的灯光。这种独特的照明方法，解决了不用灯光的直接照射而创造出衬托周围环境的光感。即将装修的房间的数码和三维打印技术将是一个观赏的安装架构和美感纯我的精品空间。

Aeds与Issey Miyake的相似点在于利用尖端的数字技术、创新的材料来保留Miyake的设计。他始终能意识到为Miyake建立展示环境根本的需要。Me 连锁服饰店作为突出的平台为Miyake的设计添加光彩。

PROJECT
Me Boutique, Issey Miyake
DESIGNER
AEDS
COMPLETION
2006
PHOTOGRAPHER
Gitty Darugar, Nomoto Masafumi

项目
三宅一生精品服饰
设计师
AEDS
完工时间
2006
摄影师
Gitty Darugar, Nomoto Masafumi

me

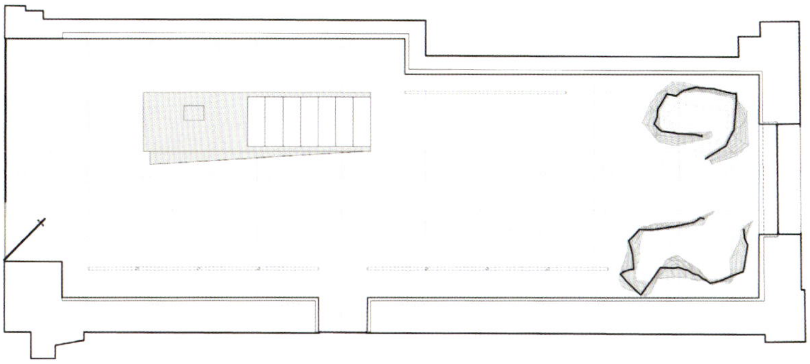

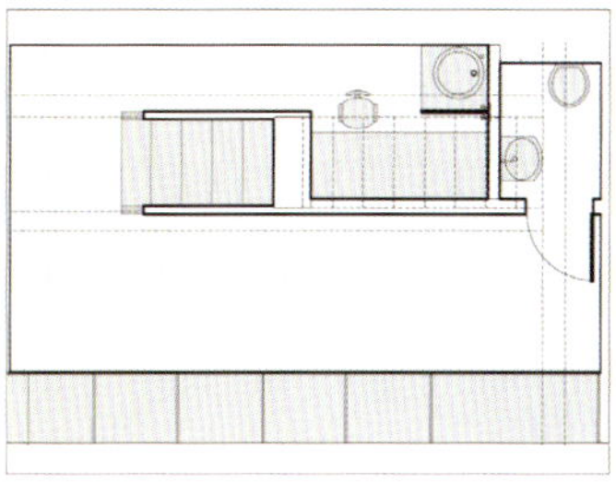

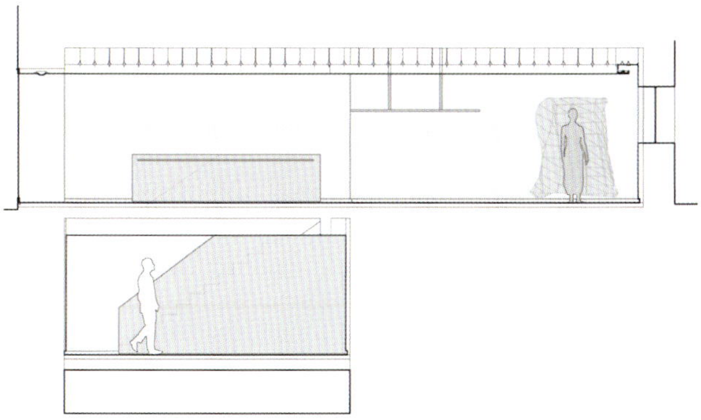

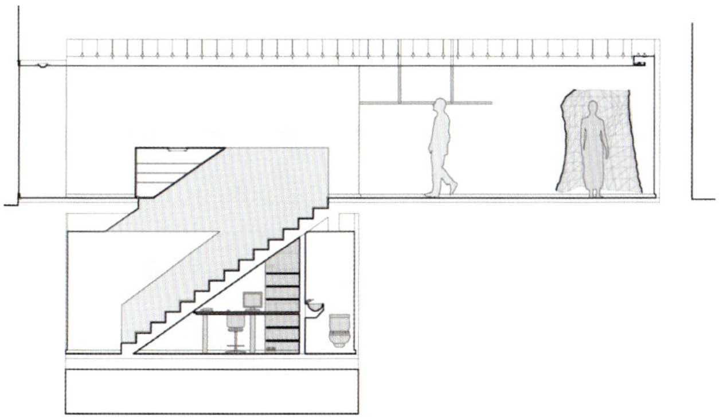

Metal House

Taipei, China

金属屋

中国，台北

A simple but strong visual image of a truly modern shopping window is created through the white coloured, exposed metal external wall. Track lighting is designed as an exhibition guide for visitors to feel the changes of the geometric layout formed with flowing lines and angular curves. From the reflection of the mirrors that wrap the columns, horizontally aligned shelves of metal corrugated boards display luxury accessories, and enliven space. Together with over 300 pieces of laser-cut punched metal webs in the ceiling, the lighting, air-conditioning and hanging displays combine in clarity and simplicity to display the clean, smooth style of digital technology. In addition to the hanging metal webbed ceiling, the cashier counter and storeroom behind the curtain address functional needs and allow more space for imagination. The streamlined curtains and display cabinets connote a tenderness of space and soften the harshness of the marble flooring and display stands without sacrificing the geometric feeling of modern design. The natural tones of the marble flooring correspond to the rhythm of the ceiling, and a premier display space is thus presented in both delicate and precise design using industrial materials.

全白色铁板喷漆的店面外墙，现代感十足的橱窗，创造出简单强烈的视觉意象。空间中活动式轨道灯引领参观视线，在全开放式的空间中，感受流线与棱角的几何形体变化，金属浪板水平层架，陈列精品配件，在经过镜面包覆的柱子反射下，使得空间更具趣味。300多片大小不同的激光切割冲孔铁网天花包覆，清爽简单及融合照明、空调、吊挂展示等功能于其中，流畅的线条展现数码科技的利落风采。天花铁网的吊挂功能、布帘后的收银台与储藏室，既满足了功能性又给人以更多的想象空间。流线造型的布帘与展示柜，刻划出柔性的空间，原本坚硬的磨石子地板与展示台都因而形成软化，但仍具几何感的形态。自然风味的磨石子与天花律动相呼应，工业素材经过精密设计呈现出高质感的展示空间。

PROJECT
Metal House
DESIGNER
Shichieh Lu
COMPLETION
2008
PHOTOGRAPHER
Marc Gerritsen

项目
金属屋
设计师
陆希杰
完工时间
2008
摄影师
马克·盖瑞森

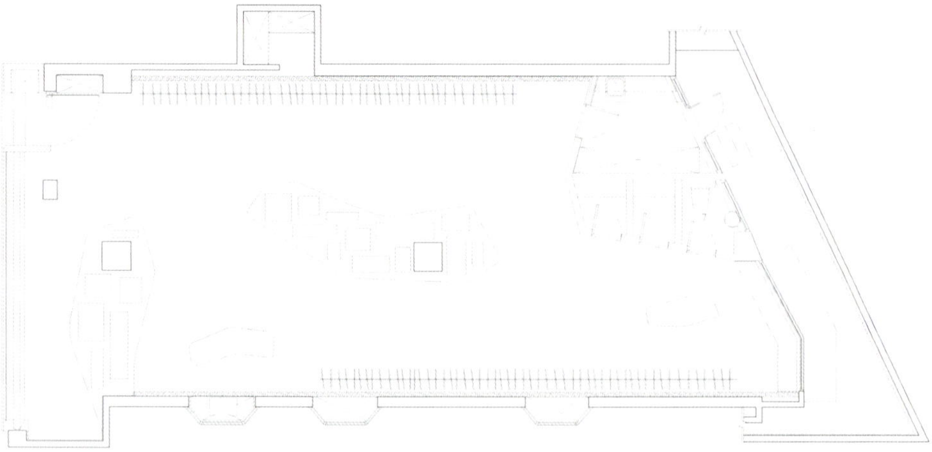

Milano Store
Milan, Italy

米兰家居店
意大利，米兰

"Corian® Design – Milano Store" is the name of the showroom opened by DuPont in downtown Milan to further increase awareness and accessibility to professionals and consumers on DuPont™ Corian® and DuPont™ Zodiaq®, two high-performance surfaces for interior design and architecture. The showroom has an area of about 300 square metres and exhibits a variety of products and systems – kitchens, bathroom decor, tables, lamps – made with Corian® solid surfaces and DuPont™ Zodiaq® quartz surfaces by client companies of DuPont, as well as installations created by designers from different countries.

The "Corian® Design – Milano Store" applies several materials and products by DuPont in its design and decor. Among these, Zodiaq® (cut in large size tiles) for the showroom floor; Corian®, for a large, backlit and curved panel (about 21 meters long by 1 meter high with surface decoration via sublimation and carving), mounted at a height of about 2.5 meters, as decorative and display covering for some walls and built-in furniture and for the three internal and external store window frames; laminated safety glass SentryGlas® Plus for the three store windows; decorative safety glass SentryGlas® Expressions™ for an interior glass wall.

杜邦公司在米兰市中心新开了一家"可丽耐®设计–米兰店"展示厅，令专业人士和消费者对杜邦™可丽耐®和杜邦™ Zodiaq®一流的内部设计与建筑有了进一步的认识。展示厅面积约300平方米，陈列着由可丽耐®设计及杜邦™ Zodiaq® 提供的石英制产品和众多国际设计大师设计的生活装置，包括厨房、浴室装饰品、桌椅、灯具等。

"可丽耐®设计–米兰店"在设计和装饰上采用了杜邦公司的多种材料和产品。其中，包括Zodiaq ® 为展示厅的地板设计（大尺寸瓷砖的切割）；可丽耐®的大型背光镶嵌板（约21米长，1米高，表面经过升华和雕刻处理），高度约2.5米，作为装饰的同时，展示一些墙壁和内置家具以及3个内部和外部存储窗框；Sentryglas ®为3家店面窗户提供的夹层安全玻璃；Sentryglas ® Expressions ™为一室内的玻璃墙提供的装饰安全玻璃。

PROJECT
Milano Store
DESIGNER
Lorenz*Kaz Studio
COMPLETION
2007
PHOTOGRAPHER
Communication in Pix + txt

项目
米兰家具店
设计师
Lorenz*Kaz Studio
完工时间
2007
摄影师
Communication in Pix + txt

CORIAN
DESIGN

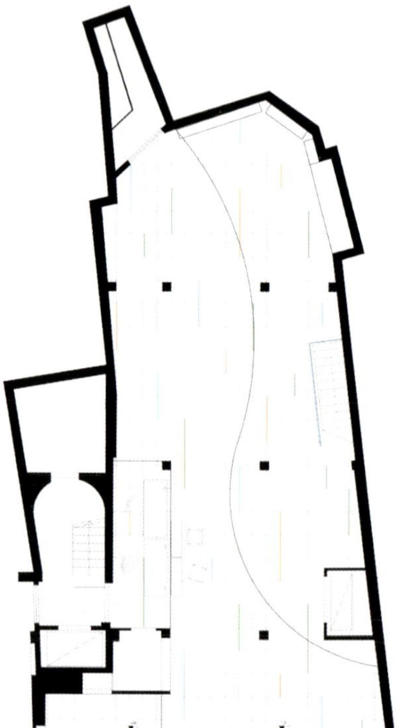

Corian
2007
40 YEARS
40 DESIGNERS
1967

Miss Sixty/ Energie store

Catania, Italy

Miss Sixty女士精品店

意大利，卡塔尼亚

A traditional atmosphere was recreated in this space, along side the general concept for Miss Sixty and Energie brands. The designer think a creative design is a journey which share between customer and designer. Only close collaboration and understanding can lead a project to new horizons.

This space brings back to life a play of contrasts between up-to-date materials and bright colours, together with traditional elements and vintage pieces to create this particular atmosphere.

The Energie Café is an interpretation of the old fashion bars and café that still exist in this area. They were called "the conversation circle" where people met to talk about politics and football. Traditional materials together with vintage pieces create this intimate and friendly atmosphere.

设计师认为创造性的设计是与客户分享的旅程。只有紧密的合作和理解，才能使项目达到一种新的境界。

在现代的材料和明亮的颜色中带来空间生活的反差，再加上传统的元素和经典作品，创造了这个特别的氛围。

活力咖啡厅诠释旧时的酒吧和咖啡馆仍然存在在这里。它们被称为“会话圈”，这是人们谈论政策和足球的地方。传统材料连同经典作品，创造出亲密和友好的气氛。

PROJECT
Miss Sixty/ Energie store
DESIGNER
Studio 63 Architecture + Design
COMPLETION
2003
PHOTOGRAPHER
Yael Pincus

项目
Miss Sixty女士精品店
设计师
Studio 63 Architecture + Design
完工时间
2003
摄影师
Yael Pincus

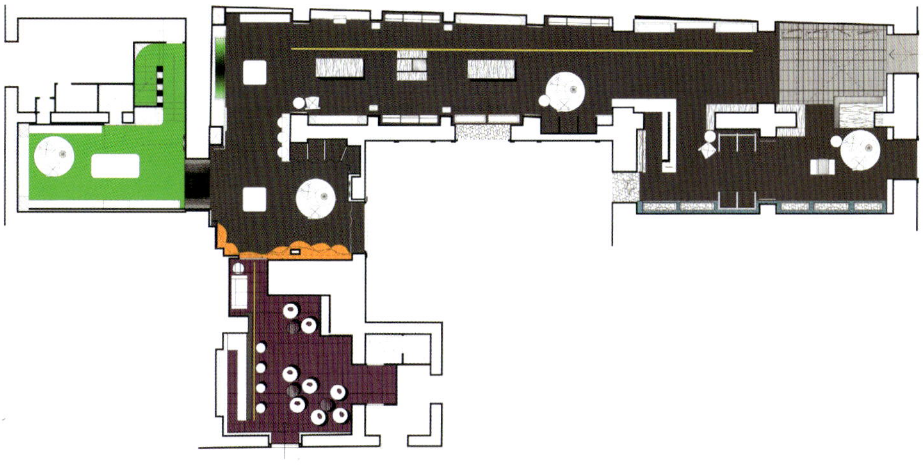

Miss Sixty Milano
Milan, Italy

Miss Sixty女士精品店
意大利，米兰

Each shop of the same brand is different from place to place and has its specific taste and atmosphere. "Each project tells a story" the designer said. Each project is a new opportunity for growth and development.

Here you can find a glamorous world of fantastic scenery and an imaginary reality. An organic space designed with curved walls,sculptured counters, soft and cozy surfaces with bright colours, all are made to create a reality bigger than life. A space gives you the freedom to be who ever you want to be, the freedom to be yourself. In the heart of fashionable Milan, giant flowers are lit up; precious materials such as ceramic tiles and luscious moquette cover walls and ceiling. Referenses to Andie Warhole pop art and the 70s' elememts become a giant 3D flower that lighst up.

同样品牌的不同店铺，风格都不尽相同，都有其特殊的味道和氛围。"每个项目都讲述了一个故事"，设计师如是说。每个项目对设计师来说都是一次成长和发展的新机会。

在这里，您可以找到一个迷人的世界，神奇的风景和一个假想的现实。弧形墙壁与有机的设计空间，条纹状的柜台，柔和舒适的表面，所有的创作都超出了现实生活。给予你随意改变自己的自由遐想空间。在米兰的时尚心脏，巨型鲜花灯光，珍贵的材料，如陶瓷地砖和包括墙壁和顶部的华美的毛毯。归诸于Andie Warhole的流行艺术和20世纪70年代演变成的3D巨型鲜花灯光。

PROJECT
Miss Sixty Milano
DESIGNER
Studio 63 Architecture + Design
COMPLETION
2003
PHOTOGRAPHER
Yael Pincus

项目
Miss Sixty女士精品店
设计师
Studio 63 Architecture + Design
完工时间
2003
摄影师
Yael Pincus

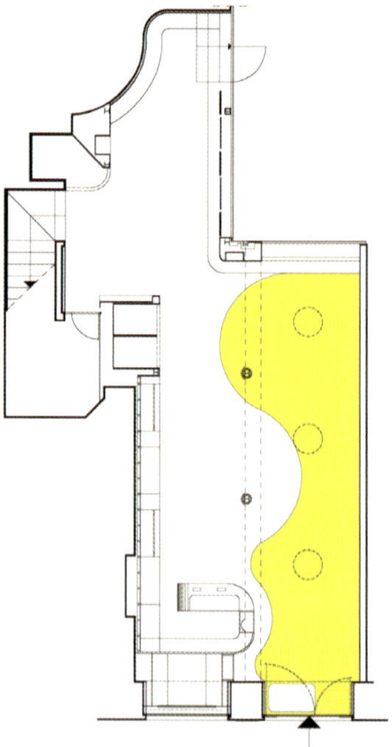

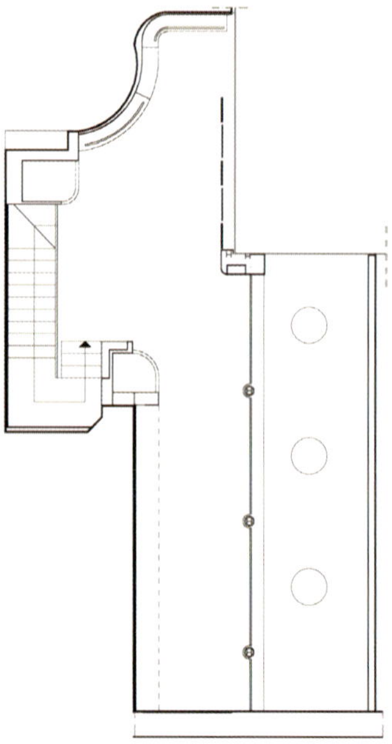

ISSEY MIYAKE
TANINO CRISCI
GARDENIA RESTAURANT CAFE
FRETTE

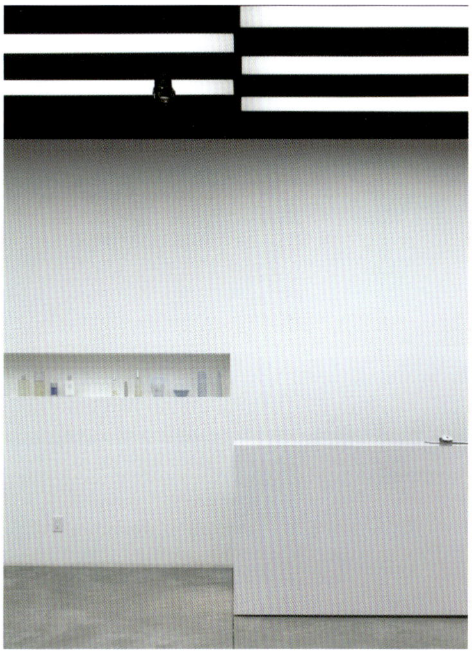

MIYAKE MADISON

New York, USA

三宅一生麦迪逊店

美国，纽约

This project for the Japanese fashion designer Issey Miyake on Manhattan's Madison Avenue consisted of an interior renovation and a new façade on this historic shopping street. The design breaks the threshold between the sidewalk and the world of Issey Miyake. The flush clear glass façade creates a seamless barrier from which the volume of the shop is extruded. The 2,600 square foot volume which terminates in a mirrored wall extends without end. In effect, the entire shop becomes a giant store window with mannequins dispersed throughout the entire depth of the space.

I-section steel columns inherited from the 1960s residential building above are the only interruptions to the volume. As the columns dance through the space with slight misalignments, they appear casual. A stretched checker-board pattern of light boxes in the ceiling meanders around these columns adding a figural element to the volume. The clothing is organized along one long wall and gives the strongest splash of expression to anotherwise understated white washed space allowing the collection to be the focal point of the experience. The façade frame is clad in black Corian with a milled debossed logo of the designer. The project had a four month schedule for design and construction and a $400,000 budget.

这是日本时装设计师Issey Miyake 在曼哈顿的麦迪逊大街这个历史悠久步行购物街店面的内部革新的一个项目。设计打破了Issey Miyake世界与街道的界限，与地面平齐的玻璃门面无缝结合，体现了商店体积的延伸。241.5平方米的面积像没有尽头的镜面墙一样无限地延伸。实际上，整个商店变成一个巨大的零售窗，模特分散在整个深度的空间。

一节节钢柱继承了1960年代上部的住宅楼面，节节钢柱像舞蹈似的轻微的偏差通过空间。一个延伸的小方格棋盘形图案蜿蜒的围绕空间中的柱体变成了一个形象的面积结构。服装是有组织的顺着墙壁延伸，并给出了最强烈的喷溅表达，白色水洗空间让收藏品成为购物的焦点。该外观框架是用黑色的可丽耐与水磨凹槽设计标识。

PROJECT
Miyake Madison
DESIGNER
Issey Miyake
COMPLETION
2006
PHOTOGRAPHER
Mikiko Kikuyama

项目
三宅一生麦迪逊店
设计师
Issey Miyake
完工时间
2006
摄影师
Mikiko Kikuyama

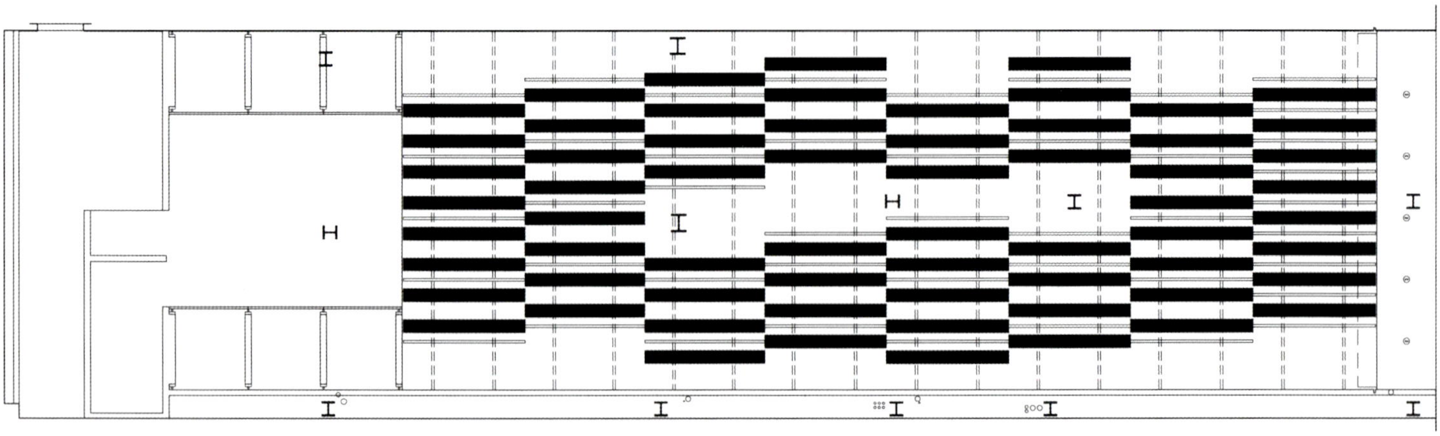

Muratti Donna
Athens, Greece

木拉蒂·唐娜店
希腊，雅典

Muratti Donna is the extension of the retail store "Muratti Uomo" at Halandri-Athens, promoting female clothes and accessories.

The organic and futuristic shapes that have been used are related to the shapes of the existing one with a lot of new elements giving the space a warmer atmosphere adapted to woman's idiosyncrasy.

The floor is covered by brown resin, matching the colour of the wallpaper and the shelves. Hangers are made of stainless steel, supported on the floor by vertical columns,helping to set the clothes in better order.

商店位于雅典，主要出售女性服饰和用品。

商店在设计上保留了其固有的结构和形状，并增添了一些新的元素，使得整个空间洋溢着女性特质的温和氛围。

地板表面是棕色的树脂涂层，与墙壁和货架的颜色和谐搭配。挂衣架由不锈钢材料制成，并由垂直的圆柱支撑固定在地面，以确保衣物保持整齐的顺序。

PROJECT
Muratti Donna
DESIGNER
Maria Lantavos
COMPLETION
2007
PHOTOGRAPHER
Antonis Karis

项目
木拉蒂·唐娜店
设计师
Maria Lantavos
完工时间
2007
摄影师
Antonis Karis

8
65
1
5
00
12
1

NIKE Soho
New York, USA

耐克Soho旗舰店
美国，纽约

21 Mercer is the new flagship store for the Nike Sportswear brand in New York's world famous SoHo district and the only location in the world where certain new Nike products are introduced and sold. The consumer experience progresses through a sequence of free-floating volumes within a larger gallery-like space.

With a material pallet ranging from blackened steel to reclaimed wood, a conscientious move toward the use of raw and recycled materials is made throughout the store. Contrasting and complementing these materials is a bright orange powder-coated room in addition to the Nike products themselves. Architectural elements from the historic cast-iron building, such as ornate Corinthian column capitals are preserved and incorporated into the vibrant new design. Changing lighting and video programs keep the space bright, colorful and engaging to the visitor. Customized bleacher and louvered walls allow the shopping environment to continually change as new products are introduced. 21 Mercer is also the home of Nike Bespoke, a new concept in custom, computer-designed footwear available by appointment only. In this workshop environment, customers work with Nike designers to build shoes to their exact specifications.

21摩丝是耐克运动品牌新的旗舰店，位于美国纽约世界闻名的Soho区。新的耐克产品在这里上市和销售。

整个商店从里到外都充分利用了环保材料：如装饰性黑色钢片和可回收木材。漆成亮橘色的房间和耐克产品与这些环保材料形成对比，同时也是对它们的补充。设计师保留了原有的一些建筑元素，如科林斯式圆柱等，并将其融入到新的设计之中。变幻的光线和视频节目使空间更加明亮、富于色彩感，也更加吸引游客。特别设计的露台百叶窗使购物环境随着新品的上市而不断变化。
21摩丝也是第一家耐克商品定制店，实现了“订购电脑设计的运动鞋”这样一个新的购物概念。在这样的购物环境中，顾客与耐克品牌设计者共同打造自己理想的商品。

PROJECT
NIKE Soho
DESIGNER
Nike Brand Design and Frederic Schwartz Architects
COMPLETION
2008
PHOTOGRAPHER
Peter Aaron, Esto

项目
耐克Soho旗舰店
设计师
耐克品牌设计和弗得里克建筑师事务所
完工时间
2008
摄影师
皮特阿隆，埃斯托

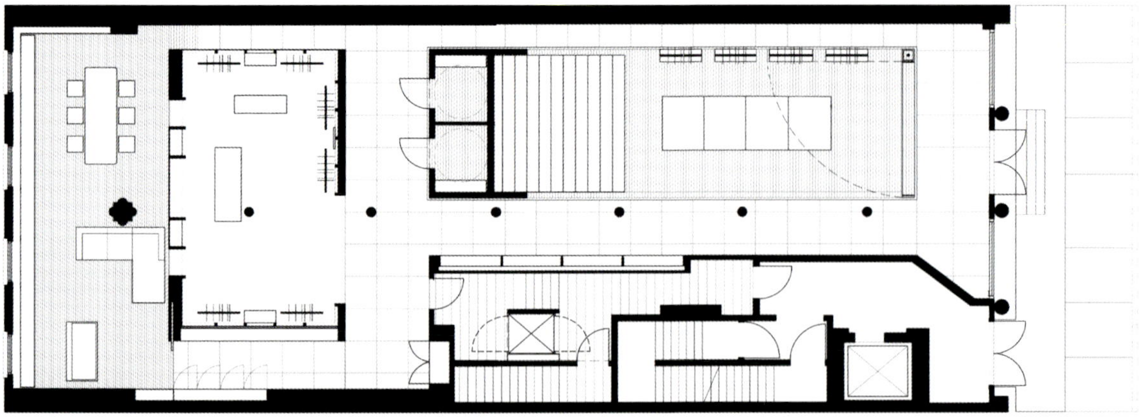

OLIJF
Tokyo, Japan

OLIJF 时尚店
日本，东京

OLIJF is a fashion select-shop targeting mature women with strong attitude.

By challenging the evolving boundary between art and design and striving to recreate the relationship between objects, space and the body, Laur's creations remain sensitive to how humans interact within defined spaces. Her methodology inevitably incorporates the exploration of colour, materials and new technologies.

For this project, Laur Meyrieux has contributed not only as an interior designer, but also as an overall art-director. She has achieved to build up its total vision, starting from the concept, which developed into its total image and choices of collection.

OLIJF是一间针对具有明确购物态度的成熟女性的时尚店。

挑战不断发展的艺术与设计之间的界限，力求重建物体、空间和人之间的关系。设计师的作品能够灵敏的感知到人与特定环境之间的相互作用。她的方法就是探寻色彩、材料和新技术之间的必然联系并使之结合。

店内设计营造了一种豪华的、但亲切的气氛。所用材料体现了购物环境的幽雅。地板的设计灵感来自艺术家画室，地板水滴状漆画装饰使得空间的形象感提升。另外，移动的照明装置模糊了空间的界限。窗户上的樱花图案更体现了这里的诗意。

PROJECT
OLIJF
DESIGNER
Laur Meyrieux
COMPLETION
2006
PHOTOGRAPHER
Laur Meyrieux Studiea

项目
OLIJF 时尚店
设计师
Laur Meyrieux
完工时间
2006
摄影师
Laur Meyrieux Studiea

alive
DESIRE
without REGRET

ATTITUDE
A MOMENT
I'd rather walk through a fire than around it
without REGRET

alive

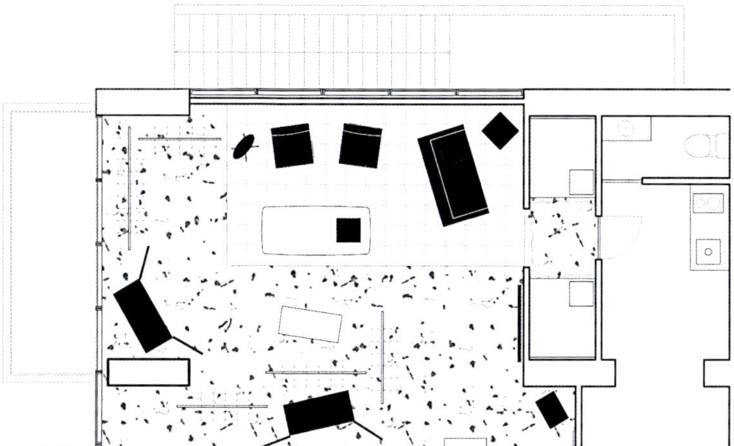

iMac

Opticon HH
Hamburg, Germany

精品眼镜店
德国，汉堡

A feeling of well being, trust, spaciousness, intimacy and flexibility are the key factors for the shop design. Rather than separating interior and exterior by a special shop window, the store stages a holistic space for display and interaction which can be read from the street when passing by. As an open space, vibrating within its boundaries, it remains free and is dominated by two large pieces of furniture. Waved areas give a dynamic to this free and open space which, with the help of mirrors, seems to continue endlessly.

The ornate design on the carpet helps carry customers through opticon's long, narrow selling space. Undulating, cleverly lit wall shelves help highlight the 2,100 pairs of glasses on display within the store. And the choice of colours, helps make the store a "strong, clear and inviting concept."

The dominant design element soft seating in the centre of the store, is an intriguing and playful design element. But at the same time, it does not overpower the space. It is innovative, yet simple. The big idea comes across.

商店的设计宗旨就是要营造一种幸福、信任的感觉，一个宽敞、亲切的空间。正是基于这样的理念，设计师并没有采用特殊的橱窗将商店的内外隔离开来，而是规划了一个整体的空间，既用于摆放货品又增加了与外界的互动交流。这是一个自由开阔的空间，波浪状的区域为其带来了一丝动感，镜子的使用则使其隐隐地一直延伸下去。

奢华的设计，狭长的售卖空间能够促使顾客长时间的驻足于Opticon。凹凸起伏的、精巧的灯光货架墙突出了店内展示的2100副眼镜。色彩的选择更使店铺展现了“浓烈、清晰及诱人”的设计理念。

突出的设计元素——商店中心的柔软座椅——是一个有趣并引人注意的亮点。但它并没有压倒整个空间，这是一个革新，也是一个简约的设计概念。

PROJECT
Opticon HH
DESIGNER
Graft
COMPLETION
2007
PHOTOGRAPHER
Christian Barz

项目
精品眼镜店
设计师
Graft
完工时间
2007
摄影师
Christian Barz

TOM FORD
PRADA
PRADA

OLIVER PEOPLES
OLIVER PEOPLES

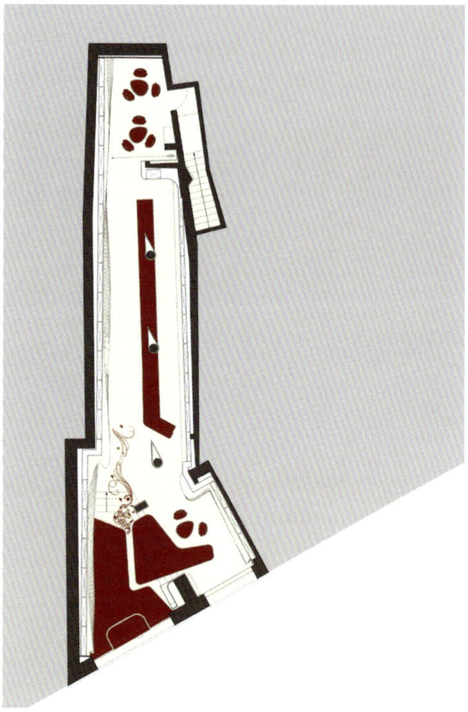

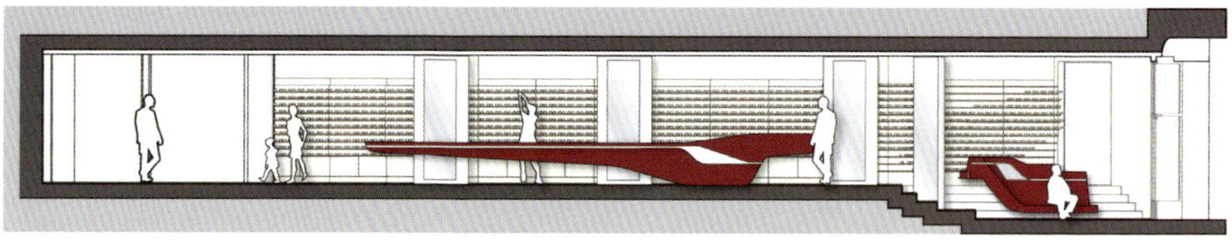

OPTICON

MYKITA

RetroSpecs & Co.

PAMPER HEiRESS
Taipei, China

宠爱后裔服装店
中国，台北

Pamper Heiress, the meaning for the successor who is in favor with, in order to echo this brand the objective, Shichieh Lu made a head concept shop for the fashion brand "Pamper Heiress" display place. He used the 'Diamond House' as the design development main idea. In this project was made for their own design clothing and jewelry to display, therefore, it transforms the diamond shape as the spatial element. In this diamond concept, the vision focus is a glass showcase crafted to imitate the multi-angle light reflections of diamonds, with various diamond shaped platforms on the four sides. The images of diamonds are everywhere, such as the magnified "diamond-ring stand" hooks in the fitting rooms. The ceiling is decorated with curved and folded lines and shapes corresponding to the angled jewelry showcase for a totally consistent look. Because of so, it is not only achieved a coordinated feeling space but also appears an interesting and direct spatial image.

Furthermore, Shichieh Lu emphasized all the materials must be complying with the main concept which is bright feeling as diamond. The whole area design is neat and smooth from large items to fine details, creating a sophisticated space just like diamond.

Pamper Heiress的意思是受宠爱的继承者。为了呼应此品牌的宗旨，设计师陆希杰为服装品牌Pamper Heiress打造的首间概念店，以"Diamond House"做为设计的主题。本商店同时陈列自创服饰与珠宝，因此将钻石造型设计为空间的元素，用玻璃切割拼接呈现出钻石的多角反射，使珠宝展示柜成为空间的视觉焦点。四周运用钻石的不同切割法打造出玻璃造型，更衣室的挂衣钩更是放大版的钻石戒座。天花板为了与角状的珠宝柜台搭配，运用折迭弯曲的线条，不仅达到整体空间的协调感，也延伸出有趣而流畅的空间意象。

除此之外，陆希杰强调在材料的筛选上必须符合主要的概念，从空间至细节都仔细处理，充分营造出钻石精致质感。

PROJECT
PAMPER HEiRESS
DESIGNER
Shichieh Lu
Ben Tsao
Athena Liu
Chia Cheng
COMPLETION
2008
PHOTOGRAPHER
Marc Gerritsen

项目
宠爱后裔服装店
设计师
陆希杰
曹车政
刘致佑
郑家皓
完工时间
2008
摄影师
马克·杰瑞提森

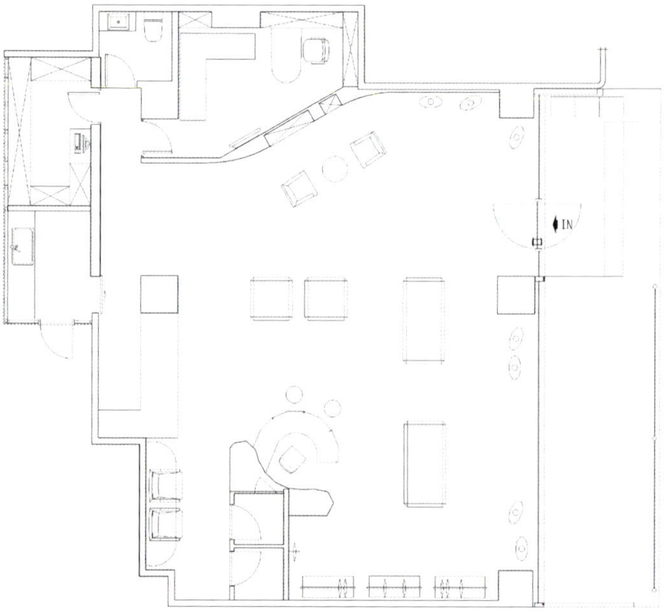
IN

Campagnole

Delicatessen shop OLIVINO

London, UK

美酒冷餐店OLIVINO

英国,伦敦

OLIVINO is a delicatessen shop complementary to the newly open restaurant OLIVOMARE, with which it shares the appealing aubergine coloured shop front as well as a graphic taste for its interior design.

Of rather small proportions – if one only considers the part of it accessible to customers on its left side, as to the entrance, a staircase protected by a full height frameless glazed partition leads to the storage in the basement and is adjacent to a perimetrical wall entirely covered by a cladding finished with a double layer (white and black) of thick opaque laminated plastic on which has been engraved a decorative pattern of variously oriented bottles and glasses, which highlights the predominantly sold product in this shop, i.e. wine.

On the opposite side a system of overhanging shelves - laid out on the wall surface according to a labyrinth geometrical scheme. This last element, while alluding to the traditional cheesemonger shop because of its "chopping board" look, with its natural appearance also counterbalances the algid uprightness of all other surrounding materials.

A heavy-duty stainless steel made back counter with back lit glazing and shelves, as well as a light purple resin floor, complement all the above described features.

OLIVINO是附属于OLIVOMARE餐厅的一家美酒冷餐店，紫红色的店面以及为室内设计增添美感的平面设计元素，都很吸引人。

左侧的服务区仅占很小的比例，服务区的左边便是出口，通往地下储藏室的楼梯设置在与屋顶相连的无框架式的玻璃后面，楼梯旁边是由双层（白色和黑色）不透明的塑料薄板做制成的墙面，上面用瓶子图案进行装饰，正强调了该商店的特色商品——葡萄酒。

对面墙面上有一组悬挂式的架子，是依据迷宫式的几何而图案设计的。另外，它同时也是一家传统的干酪店，这从它的案板可以看出来，其自然的外观与周围材料的利落风格产生一种平衡的效果。

柜台由结实的不锈钢材质制成，后面是玻璃板和货架，同时还有淡紫色的树脂地板，这些都与其它的设计相得益彰。

PROJECT
Delicatessen shop OLIVINO
DESIGNER
Architetto Pierluigi Piu
COMPLETION
2007
PHOTOGRAPHER
Pierluigi Piu

项目
美酒冷餐店OLIVINO
设计师
皮耶路易吉・皮佑
完工时间
2007
摄影师
皮耶路易吉・皮佑

onivilo
MAESTRO

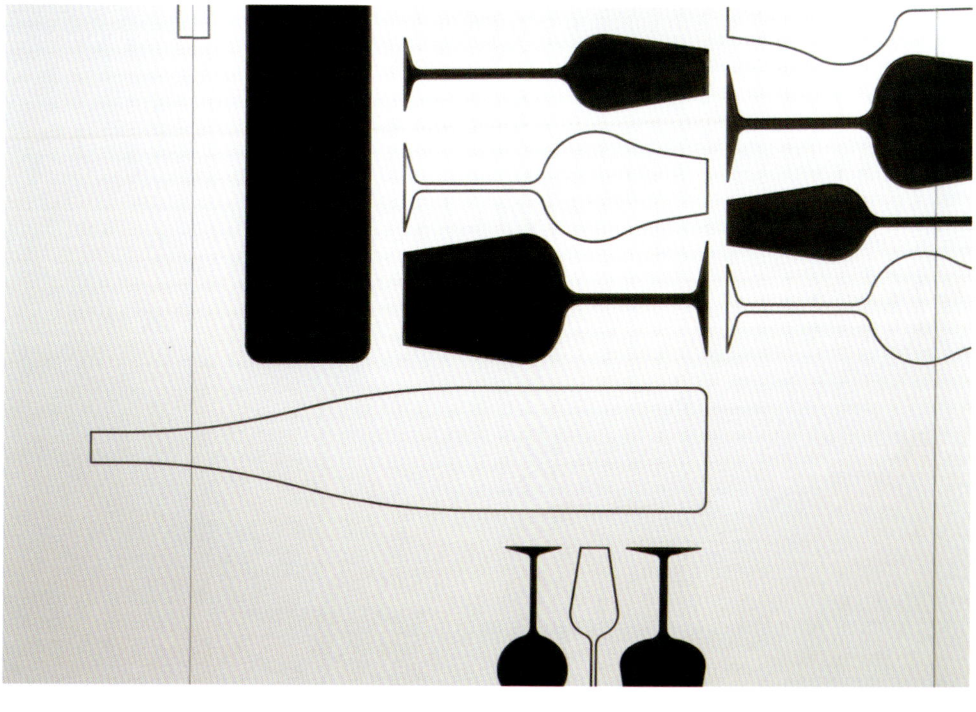

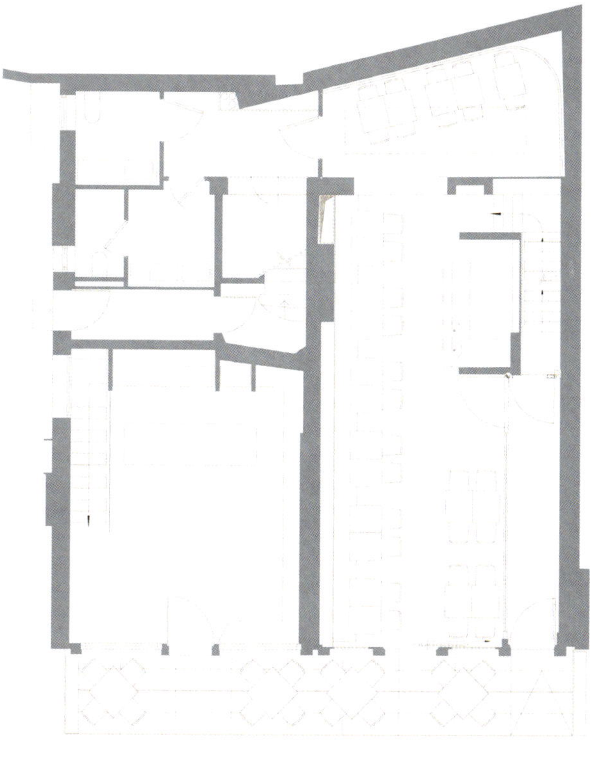

Rosato

Rosato Florence

Florence, Italy

佛罗伦萨商店

意大利，佛罗伦萨

The concept for the Rosato Retail Design Stores derives from the brand's slogan: Gold is Glam.

The designer wanted the space to resemble a gold mine, a very luxurious, glamorous gold mine. The white ceramic walls have an organic form as if handmade. On their surface are organic shaped 'holes' carved out to form the display window. The rest of the walls and ceiling are made of gold leaves, the floor is covered with off white colour moquette, the lighting is very soft and elegant, all lighting fixtures are hidden. All the materials used are precious materials that reflect the use of gold, silver and diamonds in the jewelry. Some of the products have a sense of one of a kind or hand made products and so does the space that seems to be molded by hand. The space is very clean and glamorous.

本项目的设计灵感来源于它的品牌口号：金子是光芒四射的！

我们想设计出一个类似于金矿的空间，其间让人感觉奢侈豪华，金碧辉煌。纯白色的陶瓷墙壁让人看起来像手工制成的一样。墙壁的表面布满"洞"——构成了新颖别致的展览窗。墙壁和屋顶的其他位置由金叶构成，地面覆盖有灰白的绒毛地毯，灯光柔和而幽雅，并且所有的照明器材都处在隐蔽的位置。昂贵的装饰材料与这里的黄金、白银以及各类钻石交相辉映。一些产品会给人一种手工制造的感觉，包括整个空间都让人看起来像是手工构筑而成。整个房间洁净幽雅又不乏奢美华贵！

PROJECT
Rosato Florence
DESIGNER
Studio 63 Architecture + Design
COMPLETION
2007
PHOTOGRAPHER
Yael Pincus

项目
佛罗伦萨商店
设计师
63建筑设计工作室
完工时间
2007
摄影师
雅尔·宾克斯

Rosato

Rosato

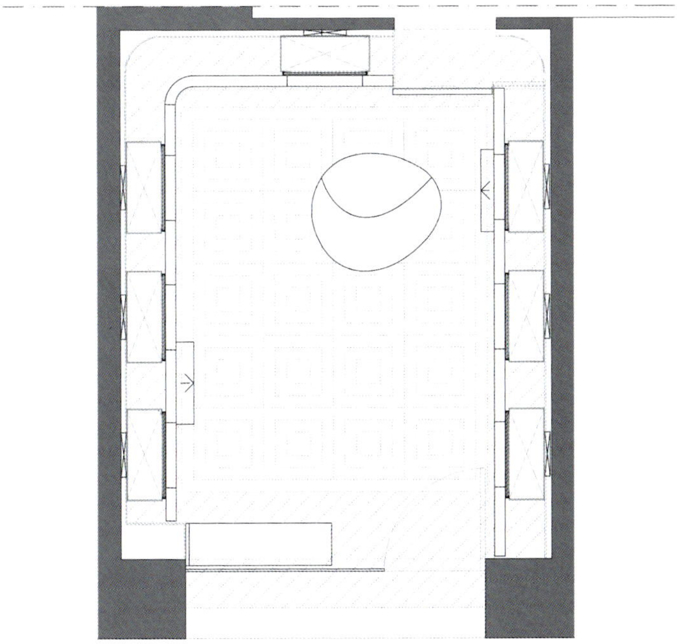

sergio rossi
EXIT

Sergio Rossi
New York, USA

赛乔·罗斯
美国，纽约

All Sergio Rossi stores are unified through a consistent design that represents the core qualities of the Sergio Rossi brand: contemporary, polished luxury.

Located in a 1,000-square-foot space on Fifth Avenue at 55th Street, Sergio Rossi is a high-end designer shoe boutique that combines the warmth of Italian luxury with the cool of modern form. The main space is enclosed by a cascading metal chain curtain that ripples like a waterfall as light softly flows down it. Suspended from this curtain wall is a high-gloss white display case which serves as a focal point for the space and highlights its modern viewpoint. A long horizontal niche is cut into this lacquered wood rectangle, serving as display space for the shoes. Free-standing teak displays positioned in front of the curtain wall provide a sense of warmth and, while grounded, give the appearance of furniture that can be moved around the space.

所有的赛乔·罗斯商店都采用了统一设计风格，以此来彰显其品牌的核心特色：现代、优雅、奢华。

这个商店坐落在第55街15大道的一个92.9平方米的面积上。赛乔·罗斯是一个高端设计师设计的鞋与意大利热情相结合的高端现代cool店。主要的空间体现在连串的叠起的金属铁窗像瀑布一样下落。悬挂的高光泽鹅白色展示面是这个空间精彩的焦点。长横向木板切入一个漆木长方形，用做展示商品的展台，可随意摆放的木质展台在金属铁窗前给人一种热情的感觉，家具也可以随意摆放在空间中的任何位置。

PROJECT
Sergio Rossi
DESIGNER
212box Architecture
COMPLETION
2006
PHOTOGRAPHER
Jody Kivort

项目
赛乔·罗斯
设计师
212box Architecture
完工时间
2006
摄影师
Jody Kivort

sergio rossi

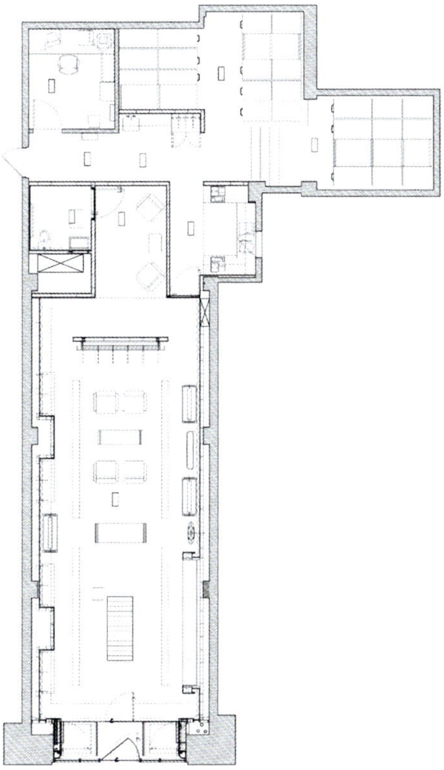

CAMPER

SIT SHOP SHU
Valetta, Malta

喜特鞋店
马耳他，瓦莱塔

The site of a new shoe shop in Valletta, a new commercial enterprise on St Lucy Street in the centre of town, presented a very limited area– one room with a basement – which provided the most significant challenge for the architects in charge of the project from the early sketches. We adopted a vertical approach rather than a horizontal one, and the volume was fragmented into one giant staircase which would combine display, seating, storage and all services into one. We imagined clients walking up, down and around the stairs, choosing their favourite shoes and actually sitting next to them to try them on, rather than looking at the merchandise through a glass. The experience of being surrounded by the products, together with the shopper and shopping in a surreal, stepped space is reminiscent of Japanese shoe-box architecture.

When it came to texture we were inspired by traditional motifs, wrought iron railings, and the Maltese eight – pointed cross traditional lace.

这家鞋店坐落在马耳他首都瓦莱塔，位于一条新商业街区St Lucy。面积很小，只有一个房间和一间地下室。设计师为了充分利用空间，巧妙地使用了楼梯。所有的功能，展示、休息、贮藏都集中在了楼梯上，我们可以想象顾客在楼梯上走来走去挑选自己喜爱的东西，走下来试穿，这样的感觉完全不同于只从玻璃窗外观看，让人不禁联想起日式的鞋盒式建筑。

古典的图案、铁制的栏杆、产自马耳他的精致蕾丝，都被运用进来。

PROJECT
SIT SHOP SHU
DESIGNER
Chris Biffa,
Bemard Vella,
Marcia Calleja
COMPLETION
2007
PHOTOGRAPHER
Chris Briffa Architects

项目
喜特鞋店
设计师
Chris Biffa,
Bemard Vella,
Marcia Calleja
完工时间
2007
摄影师
Chris Briffa Architects

CAMPER
CAMPER
imagination walks

CAMPER

CAMPER
CAMPER

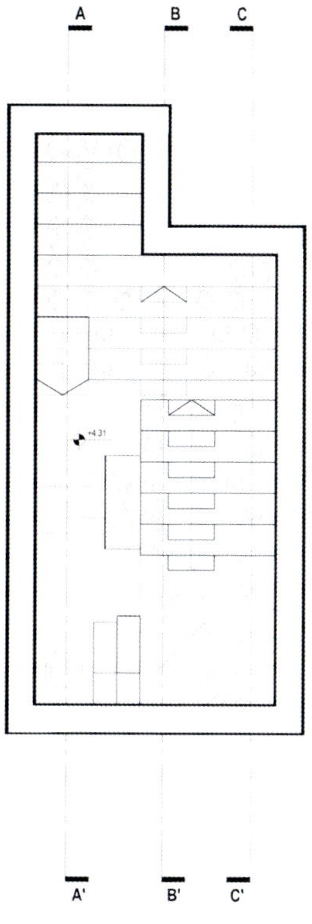
A
B
C
+4.31
A'
B'
C'

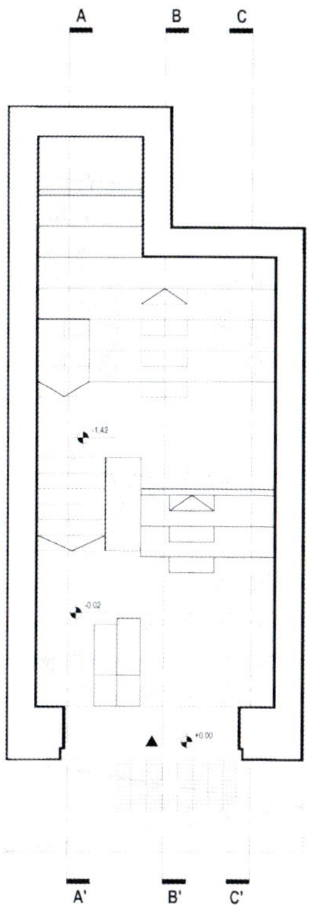
A
B
C
-1.42
-0.02
+0.00
A'
B'
C'

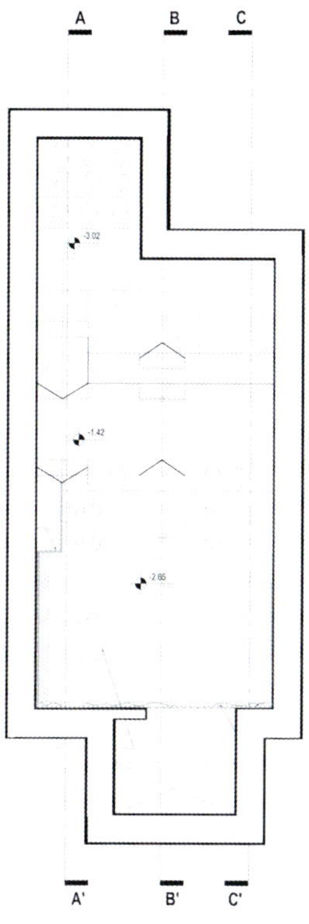
A
B
C
-3.02
-1.42
-2.65
A'
B'
C'

alifax.fr
Découvrez α

Sony-Alifax
Paris, France

索尼-阿里发克斯展厅
法国，巴黎

After 25 years of success, next to Sony, Alifax grew. it opened in October 2006, 9 rue Lafayette, a concept store next generation, combining select high-tech shop and Trade show dedicated to measure. Alifax chose the architect designer Christophe Pillet to work on the permanent demonstration of the best technology of Audio, Video, and Computer automation.

Christophe Pillet articulately analyses the ethics of his generous domain, aiming at "making life a little prettier". His style, elegant and sober, goes straight to the essential while never being stern.

Meanwhile, architect created a kind of welcoming atmosphere to make the customers feel at home and to arose their desire of purchase. Showing theprogress of modern high-tech, Sony-Alifax is no longer a sotre or showroom. The lighting, audition and structure meet together and come into a perfect combination.

在Sony成功的25年后，Alifax在不断地成长。它于2006年10月在9 rue Lafayette开设了新一代概念型展示厅。这个展示厅结合了高端技术商店和商业展示于一身。Alifax选择了设计师 Christophe Pillet 来设计这个对最高端音像及计算机自动化技术永久性的展示厅。

设计师清楚的分析了自己的设计理论和范围，旨在“使生活更美好一些”。他的设计简洁而清晰，直切主题但却不过于刻板。

设计师在设计过程中更注重了让顾客有宾至如归的感觉，不仅仅将它作为一个展示空间。使灯光、音效和布局相碰撞，实现了完美的结合。

PROJECT
Sony-Alifax
DESIGNER
christophepillet
COMPLETION
2006
PHOTOGRAPHER
christophepillet

项目
索尼-阿里发克斯展厅
设计师
christophepillet
完工时间
2006
摄影师
christophepillet

SONY
APPLE
BOSE
PIONEER

Tendo PLY
Tokyo, Japan

滕多家具店
日本，东京

Tendo Mokko is the famous furniture brand in Japan. This is the renovation of one floor of the building constructed 40 years ago. Designer makes it a space with nostalgic newness, sending respect to the accumulated time.

At using the limited budget effectively, how to design the connection of each space was an important theme and the major . Through the whole floor, the brown elements were newly added and are scattered, disposed at the important points for introducing visitor, in special, at the connection points of different spaces with priority. It also proposes the new methodology that the small things change the bigger thing or the bigger thing is renovated not by changing the whole but by changing the small part in the whole.

滕多家具是日本著名的家具品牌。这个商店是将一座40多年前建筑的一层重新翻修。设计者设计这般的怀旧气息的新空间是为了表达对旧日时光的崇敬。

利用有限的资金实现空间之间的连接，成为了这一设计主题。棕色的元素被增添到整个空间之中，零零散散地分布在重要的位置上，不仅用于引导顾客，还可以将不同的空间联系在一起。新的设计理念也被运用其中，即小的东西可以改变大的事物，这样，在整个空间不做大变动的情况下，也可以达到改变的效果。

PROJECT
Tendo PLY
DESIGNER
Daigo Ishii
Future-scape Architects
COMPLETION
2006
PHOTOGRAPHER
Future-scape Architect

项目
滕多家具店
设计师
Daigo Ishii
Future-scape Architects
完工时间
2006
摄影师
Future-scape Architect

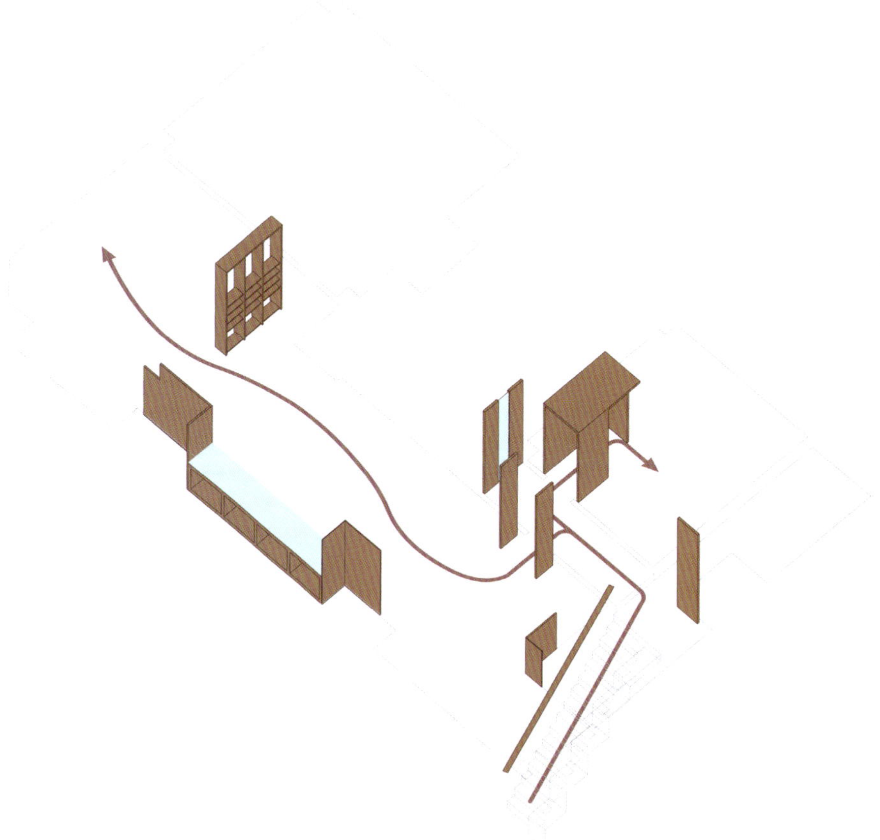

Tino Lanzi

Tino Lanzi
Sydney, Australia

提诺·蓝齐奢华店
澳大利亚，悉尼

The Tino Lanzi designs are balanced between contemporary and classical. There are shoes for the lovers of high heels, as well as lower pumps, sling back and sandals.

The soft, feminine ladies shoes store was designed to be like a modern summer garden gazebo. The walls are lined in a champagne paint finish and bronze floral wallpaper with little summer wrens and delicate ribbon stripes. The joinery is set in stark contrast using white (ANTIQUE WHITE USA) polyurethane with sharp lines creating a lattice like effect. The green carpet creates a grassy floor finish; the oak chevron floor is earthy and organic decorative motifs give formality. A huge Lily chandelier and sconces literally grow from the surfaces.

提诺蓝齐的设计平衡了现代与传统风格。这里有专为喜爱高跟鞋的人们设计的鞋子，还有一些矮跟的、露跟女鞋和凉鞋。

这是一家女士鞋店，柔美的空间如同现代的花园露台。墙壁采用香槟色的壁纸，上面饰有条形的图案和青色的花卉图案。白色细木工艺与锋利的线条打造了格子效果。绿色地毯传递清新，橡木地板彰显简约，成组的装饰图形则流露出形式感。巨大的百合花状吊灯和壁突式烛台如同从天花板上生长出来，自然和谐。

PROJECT
Tino Lanzi
DESIGNER
Greg Natale,
Stewart Horton
COMPLETION
2008
PHOTOGRAPHER
Sharrin Rees

项目
提诺·蓝齐奢华店
设计师
Greg Natale,
Stewart Horton
完工时间
2008
摄影师
Sharrin Rees

Tino Lanzi

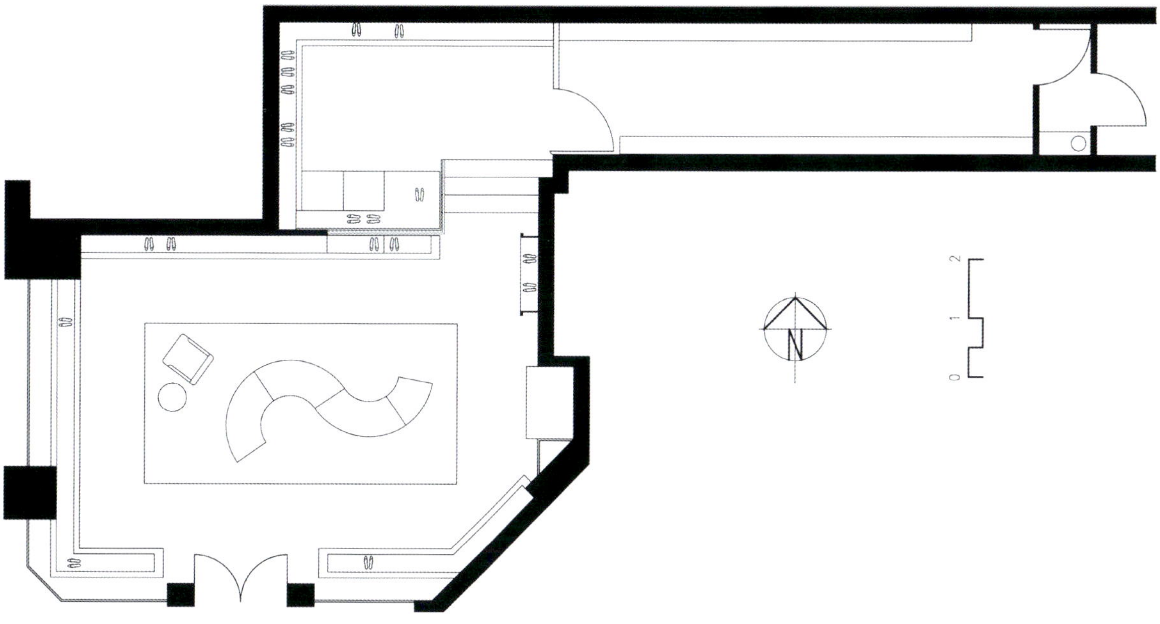
N
0
1
2

Zstore

Beijing, China

“姿”家具店

中国，北京

The designer here connects the lives with design tactfully, and makes the 8,000 square feet of show room into a fashion furniture gallery space.The two floors of the whole store are designed into square French windows, the upper of which is decorated with different brands of outdoor art posters, so that attracts the people's attention out of the store.
Walking into the store, people would be satisfied with the special display of the furniture which breaks the traditional crowded display in the furniture store.

In addition,the display shelves made form the lamp-post are attractive, displaying various of artistic seats, just like art galleries. The light platform design matches the black marble floors, giving a feeling of the stars at stage, fully demonstrates the characteristics of living art gallery.

The nearer one come into the sotre, the more surprise one will get. From the light brown wooden staircase to the wooden screen ceiling, all of the designs follow the geometric shape, which echo the square French windows in the distance.

这是一个时尚的家具概念店，在其占地约743平方米的展示空间，设计师发挥了生活和设计融合的特色，打造出仿如艺廊的时尚家具空间。

整个店面上下两层均作方块型落地玻璃窗设计，上层更以不同品牌的艺术海报作为室外玻璃窗的装饰，未进入店内已叫人自远处驻足观赏。踏进店内，宽敞的空间感有别于传统家具店的拥挤，家具的陈列令人仿佛置身舒适的家居。但更精彩的还是将墙壁设计成落地灯柱间隔的陈列架，摆放着一张张充满艺术味道的座椅，仿佛美术馆，与夹层的灯光平台设计及整室光亮的黑色云石地板互相呼应，带着一种舞台星光的味道，充分展现生活艺廊的特质。

拾阶而上，更会一层层发现设计师的匠心独运。浅啡木楼梯与木条扇屏形的天花板设计均贯穿着一分几何形态美，与店面的方块型落地玻璃窗遥相呼应。

PROJECT
Zstore
DESIGNER
Cream
COMPLETION
2007
PHOTOGRAPHER
Virgile Bertrand

项目
“姿”家具店
设计师
Cream
完工时间
2007
摄影师
Virgile Bertrand

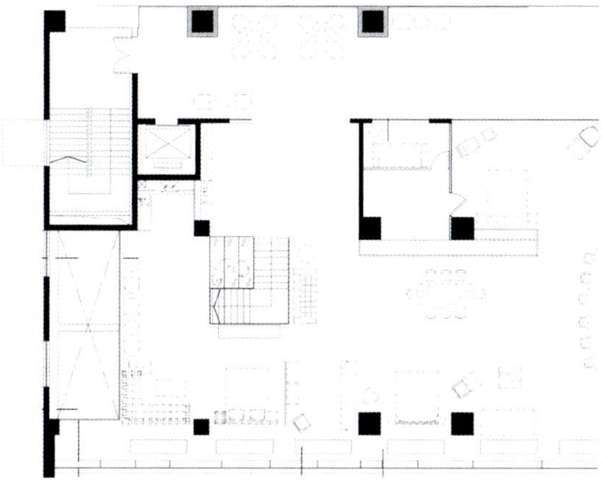

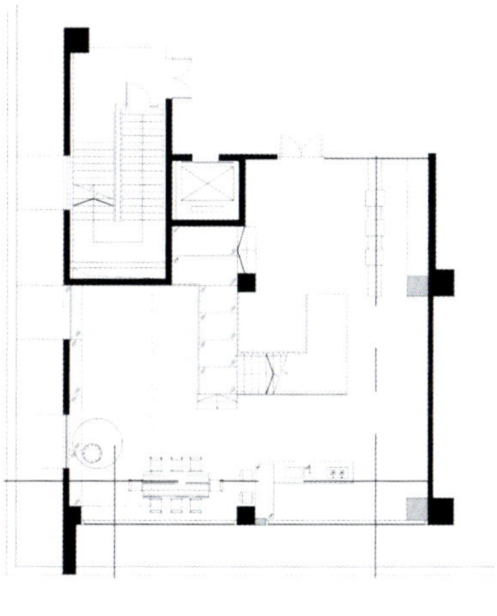

Index 索引

212box Architecture

212 box LLC, 30 Vesey Street, Floor 11
New York, 10007
T 212 233 9170
F 212 233 9168
www.212box.com

Axiom Oval Partnership Ltd

T (852) 2805-2028
F (852) 2805-2892
14/F., Malaysia Building, 50 Gloucester Road,
Wanchai, Hong Kong.
www.axiomoval.com

AEDS

5 rue de l'harmonie 75015 paris france
T 01484 24944
F 01484 24944
ammar@digit-all.net
www.digit-all.net

Bwm Architect

Bernar Walten Moser Ziviltechniker Gmbh
Margaretenplaz 4/L 1, A 1050 Wien
T 43120 59070
F 43120 59020
office@bwm.at
www.bwm.at

LHOAS & LHOAS ARCHITECTES

CH. DE FOREST 62/11 B 1060 BRUXELLES
T: 322 3720875
F: 322 3720876
www.LHOAS-LHOAS.com
info@lhoas-lhoas.com

Chris Biffa

3C Old Mint Street, Valletta VLT 1510, MALTA
T +356 2122 1407
F +356 2122 1407
chris@chrisbriffa.com
www.chrisbriffa.com

Christophepillet

29, passage dubail 75010 paris
T + 33 1 58 36 46 36
F + 33 1 42 25 01 25
caroline@christophepillet.com
www.christophepillet.com

CJ STUDIO

Floor 6 No 54, Lane 260 Kwang Fu South Road,
Taipei, Taiwan
TEL (02)2773-8366
FAX (02)27738365
cj@shi-chieh-lu.com
athena@shi-chieh-lu.com
www.shi-chieh-lu.com

C R E A M

10/F 88 Hing Fat Street
Causeway Bay Hong Kong
T +852 2147 1297
F +852 2147 0118
info@cream.com.hk
www.cream.com.hk

Daigo Ishii

Future-scape architects
Yahagi Building 401, 1-19-14, Yoyogi, Shibuya-ku,
Tokyo-to, 151-0053, Japan
T 81 3 5350 0855
F 81 3 5350 0854
info@ future-scape.co.jp
www.future-scape.co.jp

David Hicks

David Hicks Ptp.Ltd
Po Box 6110
Chapel Street North
South Yarra VIC
Australia 3141
T +61 3 9826 3955
F +62 3 9826 3988
david@davidhicks.com.au
www.davidhicks.com.au

Emmanuelle Moureaux

Emmanuelle Moureaux Architecture & Design
Tounkyo Bldg 3F, Uchikanda 1-14-14, Chiyoda-ku,
101-0047 Tokyo, Japan
T (81) 332 93 03 23
F (81)332 93 03 22
contact@emmanuelle.jp
http://www.emmanuelle.jp

Frederic Schwartz Architects

180 Varick Street New York NY 10014
212 741 3021
FSchwartz@schwartzarch.com
Phone: 212 741 3021
Fax: 212 741 2346
http://www.schwartzarch.com

G TECTS

200 hudson street 300 new york NY 10013 USA
T 2124 142300
F 2124 142301
communications@gtects.com
www.gtect.com

Giovanni D' Ambrosio

Studio Giovanni D'Ambrosio
Via Monserrato, 34
00186 - Rome ITALY
T +39-06 6869760
F +39-06 6869760
vanbergen_dambrosio@hotmail.com
www.giovannidambrosio.com

GRAFT

Gesellschaft von Architekten mbH
Heidestraße 50
10557 Berlin
T +(0)30 2404 7985
F +(0)30 2404 7987
berlin@graftlab.com
www.graftlab.com

Greg Natale

Studio 6 Level 3 35 Buckingham Street
Surry Hills NSW 2010
Australia
T +61 2 8399 2103
F + 61 2 9399 3104
info@gregnatale.com
www.gregnatale.com

Hartmut Raiser and Kristina Lopes

RaiserLopesDesigners
Hauptmannsreute 69
D - 70193 Stuttgart
T 07 11 2 48 39 19 0
F 07 11 2 48 39 19 99
Hartmut.Raiser@RaiserLopes.com
Kristina.Lopes@RaiserLopes.com

Hecker Phelan & Guthrie

1 Balmain Street
Richmond Victoria 3121
T 03 9421 1644
F 03 9421 1677
hpg@hpg.net.au
www.Hpg.net.au

Hiroshi Nakamura & Nap Architects

2-15-7-1F Sakura-Shinmachi Setagaya-ku
154-0015 Tokyo JAPAN
T +81(0)3 5426 0105
F +81(0)3 3428 0886
nakamura@nakam.info
www.nakam.info

Hiroyuki miyake

Hiroyuki miyake design office.
106, 2-2-18 Jouganji Kita – ku
Nagoya Aichi 462 – 0021 Japan
T +81 52 433 7896
F +81 52 433 7897
info@hiroyukimiyake.com
www.hiroyukimiyake.com

Horace Pan

PANORAMA International Ltd
Unit 7B, Charmill Centre,
50 Hillwood Road,Tsimshatsui,
Kowtoon, Hongkong
T 852 2317 7850
F 852 2317 7851
info@panoramahk.com
www. Panoramahk.com

Ippolito Fleitz Group

Augustenstrasse 87
70197 Stuttgart, Cermany
T +49 (0)711 993392-330
F +49 (0)711 993392-333
info@ifgroup.org
www.ifgroup.org

Keiichiro Sako

135-0062 1-9-12-1104, Shinonome, Kouto-ku, Tokyo, Japan
T +81 3 5548 4610
F +81 3 5548 4610
info@sako.co.jp
www.sako.co.jp

Laur Meyrieux

Roop Toranomon 4B/4F
2-9-2 Nishishimbashi
Minato-ku
TOKYO 105-0003 JAPAN
T +81 3 6909 3012
F +81 3 6909 3012
contact@laurmeyrieuxstudio.com

Maria Lantavos

33 TRAPEZOUNTOS STR.
AGIOS STEFANOS,ATTIKI
P.C. 14565 - GREECE
T +30 210 8834 095
F +30 210 8823 569
Contact@lp-architects.gr
www.lp-architects.gr

Peter Marino Architect

150 East 58 Street
Floor 36
New York, NY 10022
T 646-254-4974
F 212-759-3727
dmlaffite@petermarinoarchitect.com
www.petermarinoarchitect.com

Steffen Kaz and Catharina Lorenz

communication in pix + txt
via elba, 12
20144 milano - italy
T +39 3470868023
F +39 02 45491266
press@sabineschweigert.com
www.sabineschweigert.com

Studio 63 Architecture + Design

601 West 26th Street Suite 1510
New York, NY 10001
T +1 212 675 97 84
F +1 212 675 97 85
sergio.mannino@studio63usa.com
www.studio63usa.com

Yoshihiko Mamiya

LAID Bldg.3F 1-9-9,Nishihonmachi,Nishi-ku,Osaka
T 06 6110 1128
F 06 6534 0577
www.infix-design.com

图书在版编目（CIP）数据

酷店／李亮亮，胡嘉思编．—沈阳：辽宁科学技术出版社，2009.10
ISBN 978-7-5381-5778-9

I. 酷… II. ①李…②胡… III. 商店－建筑设计－世界 IV. TU247.2

中国版本图书馆CIP数据核字（2009）第007437号

出版发行：辽宁科学技术出版社
（地址：沈阳市和平区十一纬路29号　邮编：110003）
印 刷 者：利丰雅高印刷（深圳）有限公司
经 销 者：各地新华书店
幅面尺寸：225mm × 285mm
印　　张：17
字　　数：120千字
插　　页：4
印　　数：1~2000
出版时间：2009年 10 月第 1 版
印刷时间：2009年 10 月第 1 次印刷
责任编辑：陈慈良
封面设计：迟　海
版式设计：迟　海
责任校对：周　文

书　　号：ISBN 978-7-5381-5778-9
定　　价：198.00元

联系电话：024-23284360
邮购热线：024-23284502
E-mail: lkzzb@mail.lnpgc.com.cn
http://www.lnkj.com.cn
本书网址：www.lnkj.cn/uri.sh/5778